IMAGES
of Aviation

MASSACHUSETTS AVIATION

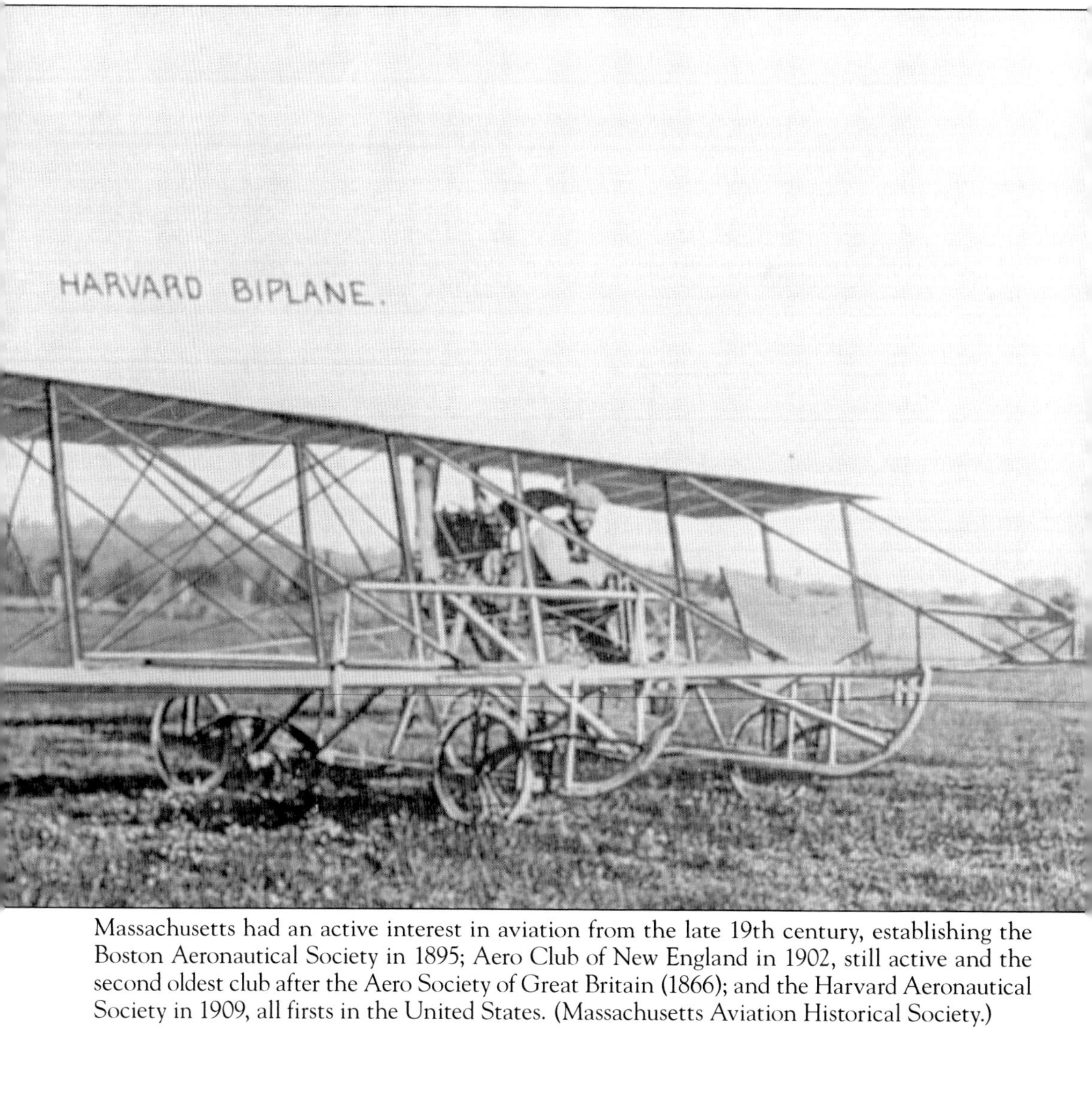

Massachusetts had an active interest in aviation from the late 19th century, establishing the Boston Aeronautical Society in 1895; Aero Club of New England in 1902, still active and the second oldest club after the Aero Society of Great Britain (1866); and the Harvard Aeronautical Society in 1909, all firsts in the United States. (Massachusetts Aviation Historical Society.)

On the Cover: Harriet Quimby dared to dream. The magazine and early screenplay writer took up flying during America's first aviation craze and flew into the history books during the 1912 Boston Aero Meet, held at Squantum in Quincy, Massachusetts. (Quincy Historical Society.)

IMAGES
of Aviation

MASSACHUSETTS AVIATION

Frederick R. Morin and John Galluzzo

ARCADIA
PUBLISHING

ISBN 978-1-4671-2402-7

Published by Arcadia Publishing
Charleston, South Carolina

Printed in the United States of America

Library of Congress Control Number: 2016936274

For all general information, please contact Arcadia Publishing:
Telephone 843-853-2070
Fax 843-853-0044
E-mail sales@arcadiapublishing.com
For customer service and orders:
Toll-Free 1-888-313-2665

Visit us on the Internet at www.arcadiapublishing.com

We dedicate this book to the early Massachusetts aviation pioneers, the men and women who dared to take to the sky despite the longest of odds.

Contents

ACKNOWLEDGMENTS

The authors would like to thank their many friends at the Massachusetts Air & Space Museum, who gave us complete access to the Massachusetts Aviation Historical Society archives; Massachusetts Aviation Historical Society (MAHS); Ed Fitzgerald, executive director of the Quincy Historical Society (QHS); the Massachusetts National Guard (MNG); Timothy Muldoon and the Massachusetts Air National Guard Historical Association (MANGHA); the US Navy (Navy); the National Archives and Records Administration (NARA); the US Coast Guard (USCG); Scott Price, acting historian, US Coast Guard; the US Air Force (USAF); Andrew Biscoe, USAF, Westover Air Force Base; Steve Rawding, Massachusetts Aeronautical Division; Ms. Danny Levy, director of strategic communications and marketing, Massachusetts Port Authority (Massport); historian Marc Frattasio; the Association of Naval Aviators Patriot Squadron; and more for their help in making this book soar to life. Thanks, too, to Lorraine and Michelle, for encouraging this joint dream to take flight.

INTRODUCTION

On December 17, 1903, Orville and Wilbur Wright demonstrated the first powered, manned flight by a heavier-than-air machine at Kitty Hawk, North Carolina. It would take just over six years for powered flight to come to Massachusetts. On February 28, 1910, Augustus Herring piloted an aircraft designed and built by Starling Burgess and Herring in Marblehead, Massachusetts, over the frozen surface of Lake Chebacco in nearby Hamilton, Massachusetts. In April 1910, Burgess established the first airfield in Massachusetts on Plum Island in Newburyport to test his designs.

The Harvard Aeronautical Society was formed in 1909 by some Harvard students, faculty, and alumni and was promoting an aero meet to be held in September 1910 at Soldiers Field in Cambridge. Almost immediately, that venue proved too confining and potentially unsafe. The society decided to move the site to the Atlantic (Squantum) section of Quincy, where the New York, New Haven & Hartford Railway Company owned a large tract of open and relatively flat land. Over the span of two and a half months the area was transformed into the Harvard Flying Field. The society's president, Abbott Lawrence Rotch, was himself an MIT graduate and president of the Blue Hills Observatory in nearby Milton. Entrants came from around the world and included the Wright brothers and their company flyers; aviation pioneer Glenn Curtiss; and Englishman A.V. Roe, later to be the supplier of the Avro bombers used by the Royal Air Force during World War II. Notable spectators included US president William Taft, Navy secretary George Von L. Meyer, Massachusetts governor Eben Sumner Draper, Boston mayor John F. Fitzgerald (grandfather of future president John F. Kennedy), and New York state senator and future US president Franklin D. Roosevelt.

World War I brought aviation to the forefront of technology, and innovations from aircraft manufacturers as well as colleges and universities developed at an extraordinary pace. At the end of World War I, returning flyers wanted to continue flying and discovered the lack of necessary resources. Small airfields and flying schools sprang up around the area. The King family in Taunton started their airfield in 1919, and today it is the oldest continuously operating airport in the state of Massachusetts. The Harvard Flying Field, which hosted the Harvard-Boston Aero Meets of 1910 and 1911 and the Boston Aero Meet of 1912, had been the site of aircraft testing for some local companies and a US Navy flight training facility. In 1923, Lt. Comdr. Richard E. Byrd convinced the Navy to construct and commission the first Naval Reserve air base in the United States at the Squantum site. Also in the 1920s, the introduction of airmail service required airfields for maintenance and emergency services. In 1927, aviation changed dramatically, and suddenly everyone wanted to learn how to fly; Charles Lindbergh flew nonstop solo across the Atlantic Ocean. (The US Navy had flown across the Atlantic in 1919, when one of three NC planes made the trip from Long Island, New York, to Lisbon, but the plane made several stops along the way; Albert C. Read, a Hanson native, was the pilot.) "Lucky Lindy's" feat greatly increased the public's appetite for flying and created a demand for flying schools and airfields throughout the state and country. As a point of interest, Lindbergh's last US landfall on his way to Paris was over the coast of Marshfield.

Throughout the 1930s, airfields sprang up across the state. Some were privately owned by people wealthy enough to afford the sport; others were established by returning veterans wishing to continue flying and to teach others. However, the Depression affected a lot of those plans, and many small airfields and flying schools disappeared. When the threat of war emerged in 1939 and 1940, plans were already underway to establish military airfields along the coast. The emergence of German U-boats along the East Coast of the United States represented a potential major threat to shipping. When war did break out in December 1941, the threat turned real, and German U-boats decimated Allied shipping along the entire East Coast. Airfields within 50 miles of the coast were closed or taken over by the military. Airfields in Plymouth, New Bedford, Mansfield, and many more across Cape Cod and the state were taken over as outlying fields (OLF) and used as training fields for Navy bases in Squantum and Quonset Point, Rhode Island. During this time, Naval Air Station South Weymouth was established as a lighter-than-air (LTA) facility for the Navy. Its giant blimp hangars would be landmarks, with one surviving until 1967.

After the war, many more flyers returned home, and general aviation again took off across the state. Returning veterans established airfields in places like Braintree and Marshfield. The OLF fields were returned to private or municipal use, and the improvements made by the Navy, like paved runways, taxiways, buildings, and lighting, were welcome additions. These fields flourished during the 1950s and 1960s as the public embraced general aviation. The commercial use of small aircraft increased significantly as well. Many small business owners used aircraft to travel on business, and agricultural enterprises embraced the efficiencies of airplanes for spraying fields. On the South Shore, small airplanes and helicopters are a common site around cranberry bogs in the spring, many flying from small grass backyard landing strips. However, economic events and the gas crisis of the late 1970s had a major impact on the cost of flying. As families grew and the rapid growth in housing pushed the suburbs further out from the cities, small airports were threatened. Then September 11, 2001, almost sounded the death knell for general aviation. New federal regulations and security issues were implemented for good cause, but the effect was to further restrict general aviation and increase the cost of pleasure and sport flying. Many general-use airports are still spread across the South Shore, and we will attempt to document these as well as the ones that have forever disappeared from the landscape

One

Pioneers

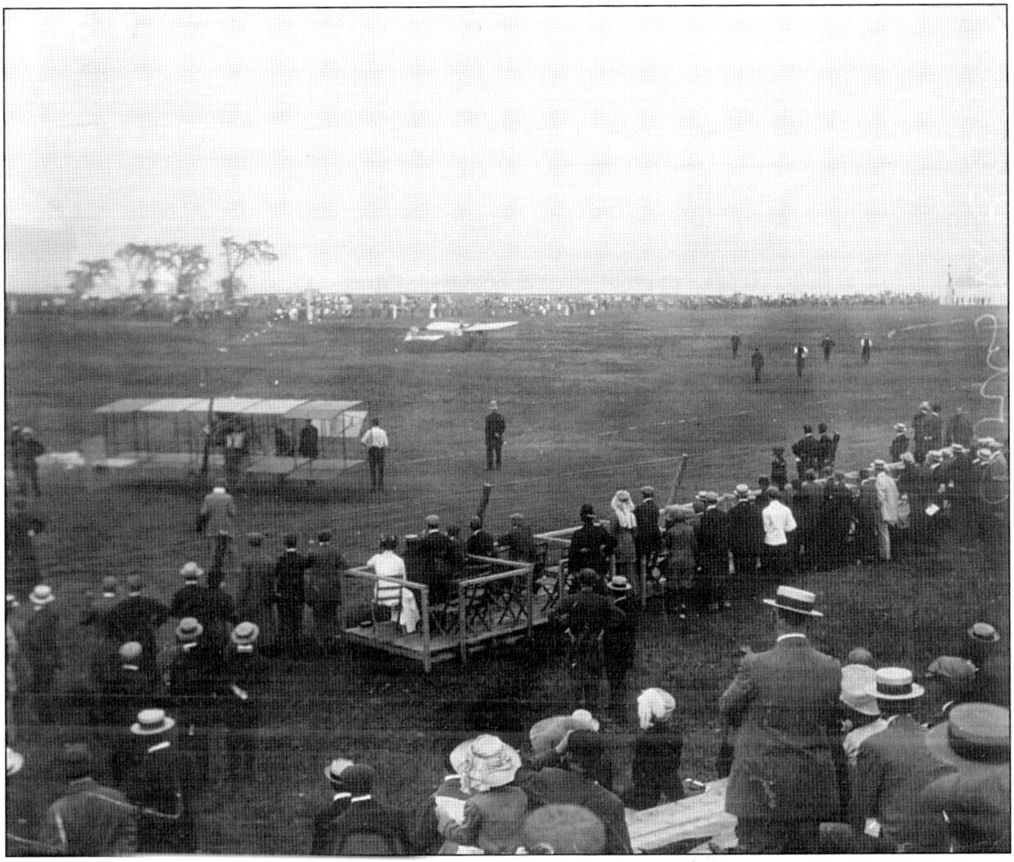

Massachusetts residents caught the flying bug early. The Harvard Aeronautical Society decided to promote an international aero meet at Soldiers Field in Boston but met with opposition from neighbors and frustration with the close confines of the field. In July 1910, the society contracted with the New York, New Haven & Hartford Railroad for property at Squantum, in Quincy, and had an airfield built in 45 days for the first Harvard-Boston Aero Meet. (QHS.)

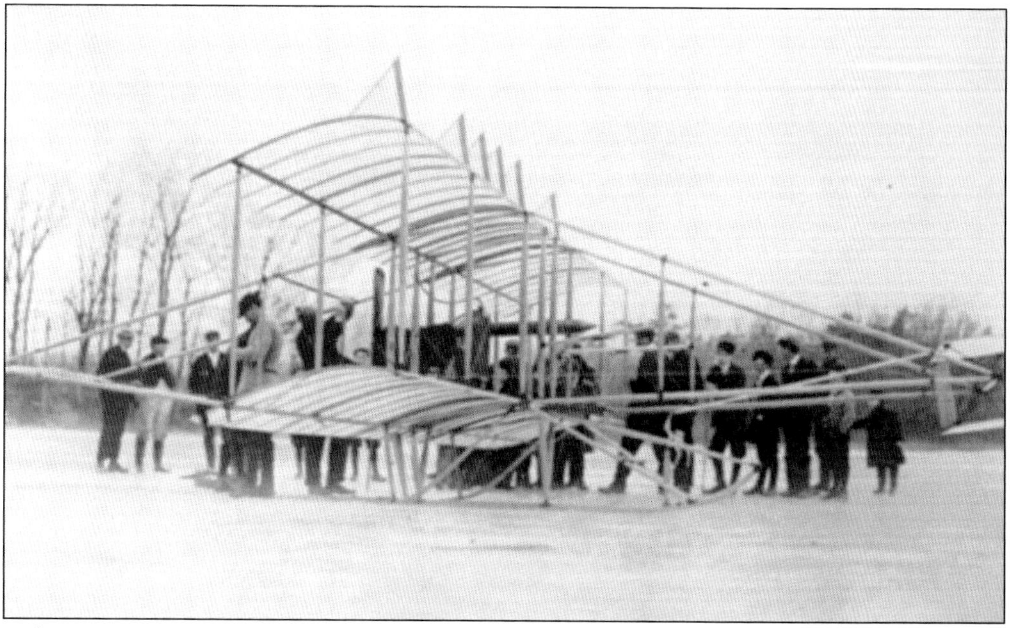

Before Squantum, though, came Starling Burgess. A successful racing yacht designer with a keen interest in airplanes, Burgess designed and built his first airplane after seeing an exhibition in Boston's Mechanic Hall. Using the family yacht design firm in Marblehead, he constructed his first plane with codesigner Augustus Herring and transported it to the Prince Estate on Lake Chebacco in Hamilton, Massachusetts. (MAHS.)

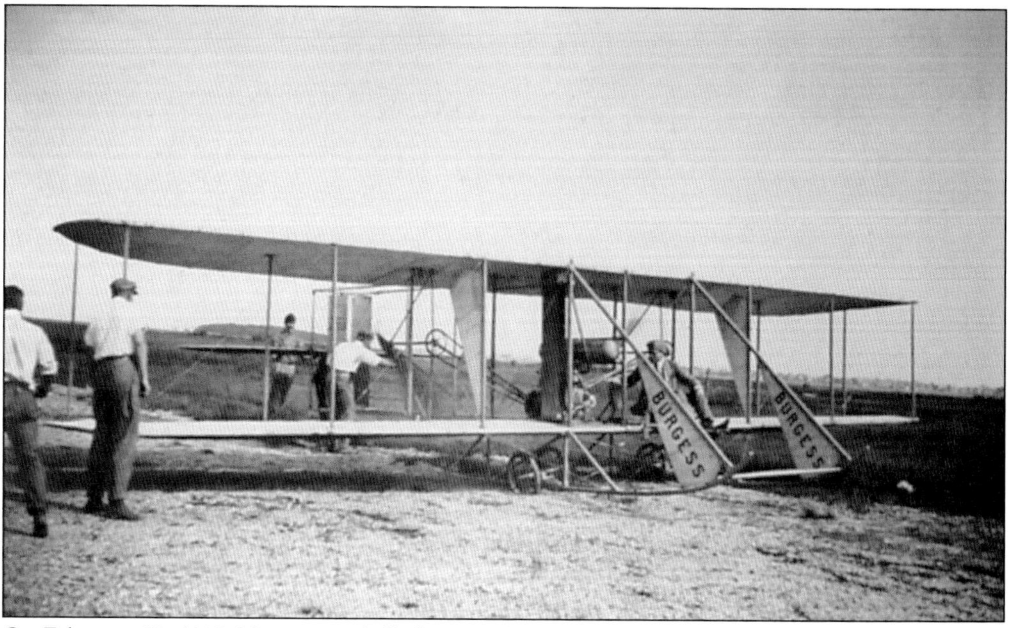

On February 28, 1910, Herring took flight over the frozen lake, thus successfully performing the first recorded flight in the state and the first in an aircraft designed and built in Massachusetts. It was also here that Burgess and Wilbur Wright concluded the first license to manufacture Wright airplanes, the Model F. Burgess is seen here at the rear of his airplane. Burgess went on to manufacture many airplanes for the US Navy. (MAHS.)

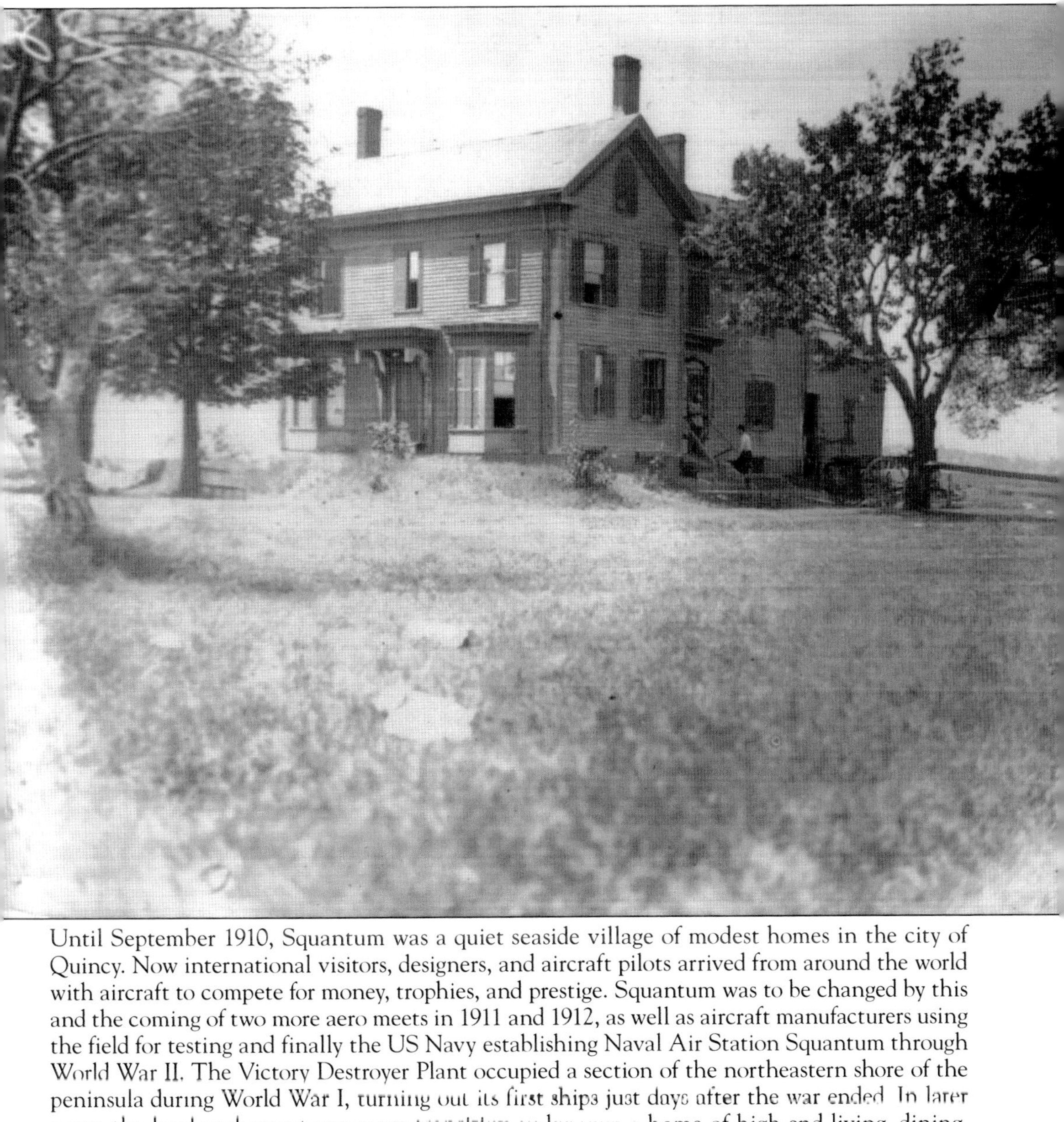

Until September 1910, Squantum was a quiet seaside village of modest homes in the city of Quincy. Now international visitors, designers, and aircraft pilots arrived from around the world with aircraft to compete for money, trophies, and prestige. Squantum was to be changed by this and the coming of two more aero meets in 1911 and 1912, as well as aircraft manufacturers using the field for testing and finally the US Navy establishing Naval Air Station Squantum through World War II. The Victory Destroyer Plant occupied a section of the northeastern shore of the peninsula during World War I, turning out its first ships just days after the war ended. In later years, the land underwent one more transition to become a home of high end living, dining, and nightlife at what is now known as Marina Bay. The sleepy farming days of the 19th century faded into oblivion the day the Harvard Aeronautical Society signed the papers to start building the airfield. (QHS.)

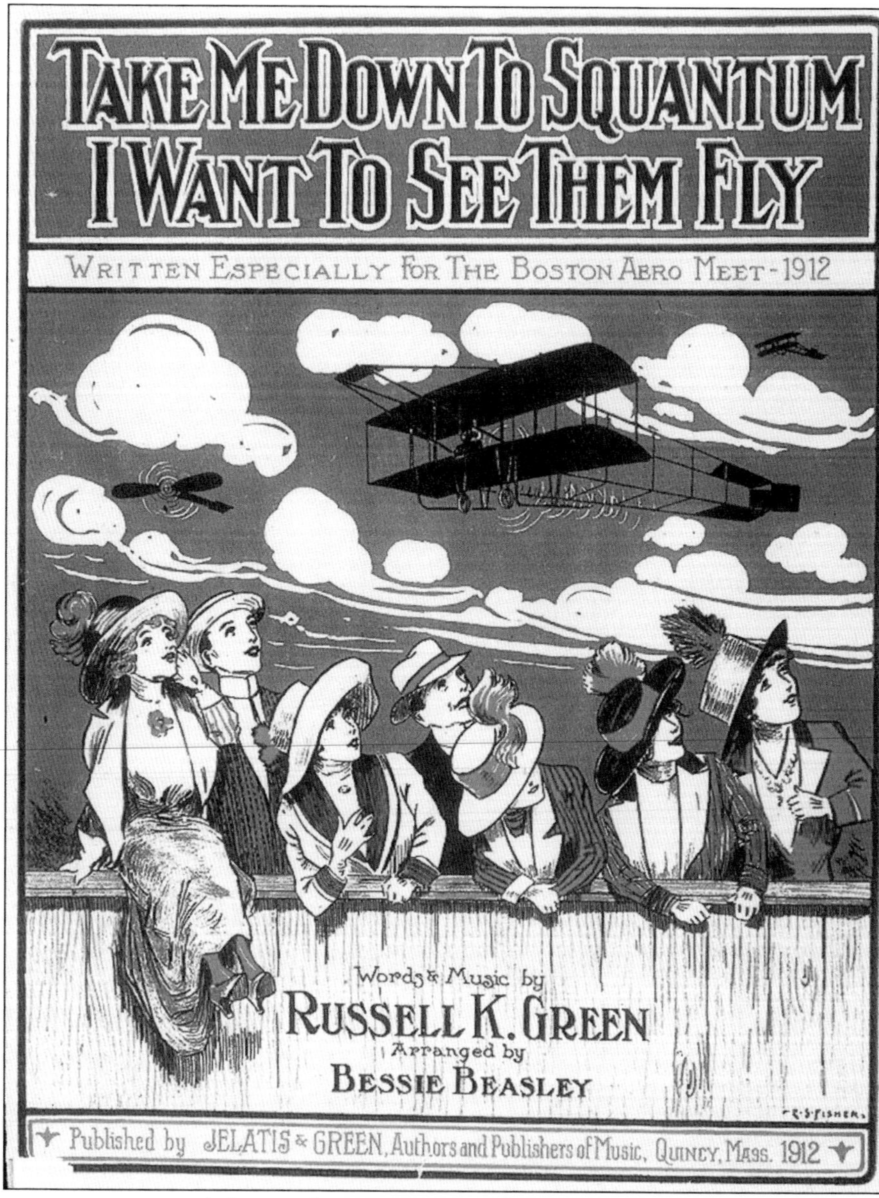

The aero meets attracted thousands anxious to see these new aircraft and marvel at their and the pilots' abilities. The 1912 aero meet even had music created to highlight the event. The 1910 Harvard-Boston Aero Meet not only drew enormous crowds locally but also attracted many politicians looking to see what this aviation thing was all about. Among those present were the mayor of Boston and future grandfather of Pres. John F. Kennedy, John "Honey Fitz" Fitzgerald; Pres. William Taft; state senator, future under-secretary of the Navy, and future president Franklin D. Roosevelt; Navy secretary George Von L. Meyer; and Massachusetts governor Eben Draper. Foreign visitors included Claude Grahame-White, the most famous flyer in the world at the time, and A.V. Roe of England, who would go on to be one of the most famous British military manufacturers of fighters and bombers in two world wars, known for his famed Avro bombers. Also in attendance was Thomas Sopwith, another British aviation pioneer, and the Russian ambassador to the United States, Baron Roman Rosen. (MAHS.)

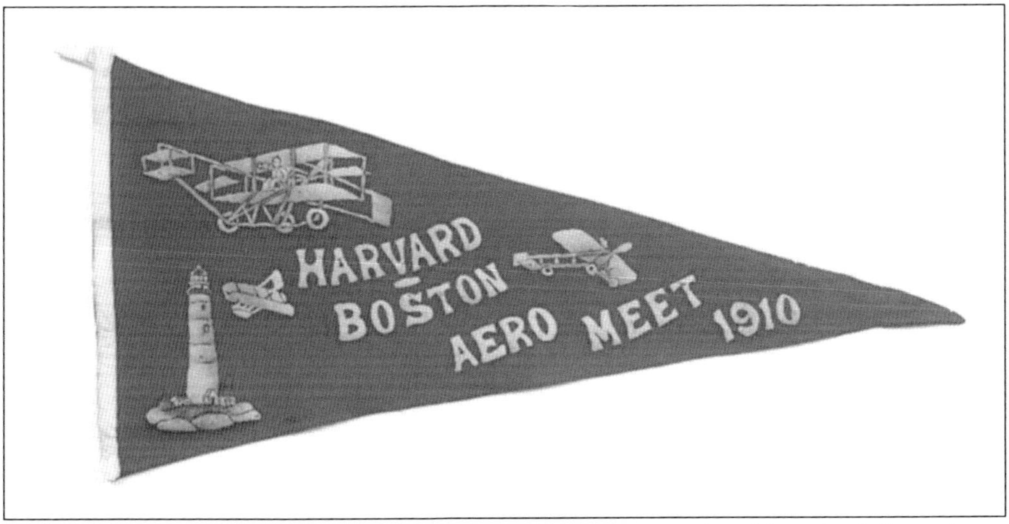

Among the most popular souvenirs at the aero meet were pennants heralding the different flying machines. Note the single-wing aircraft, as opposed to the biplane design prevalent at the time. The 1910 meet was not a financial success, nor was 1911. In 1912, Harvard dropped its association with the meet, and it was just the Boston Aero Meet. (MAHS.)

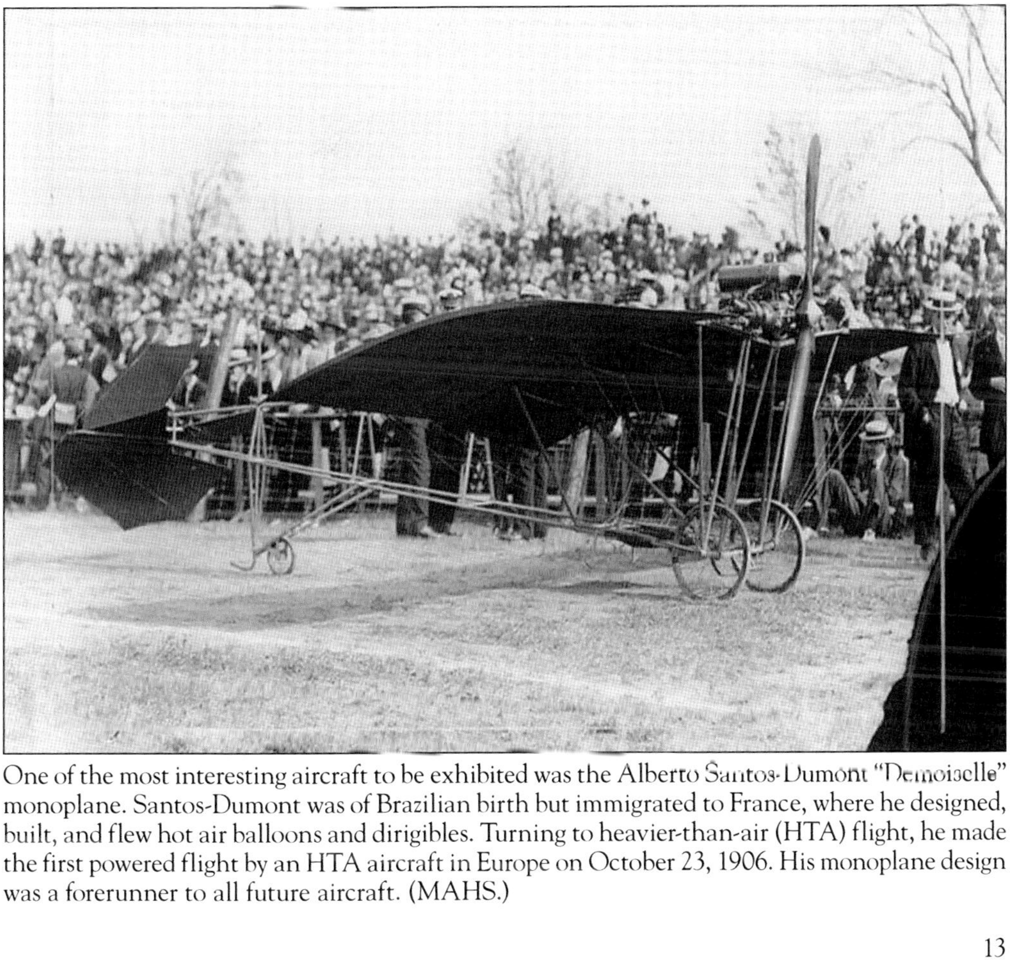

One of the most interesting aircraft to be exhibited was the Alberto Santos-Dumont "Demoiselle" monoplane. Santos-Dumont was of Brazilian birth but immigrated to France, where he designed, built, and flew hot air balloons and dirigibles. Turning to heavier-than-air (HTA) flight, he made the first powered flight by an HTA aircraft in Europe on October 23, 1906. His monoplane design was a forerunner to all future aircraft. (MAHS.)

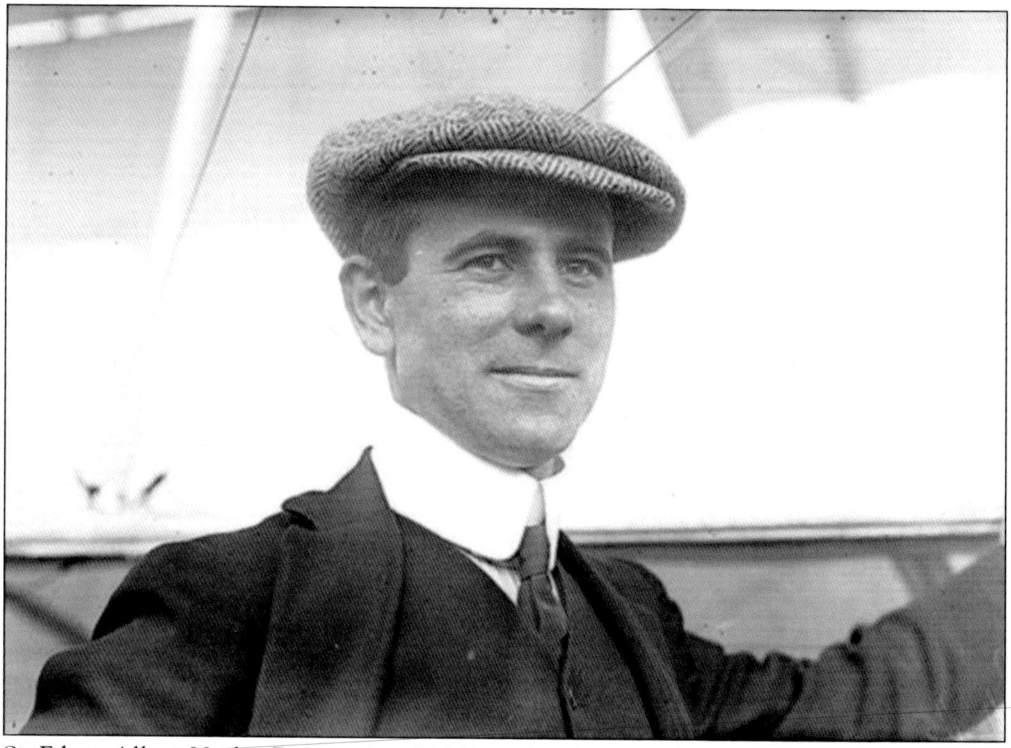

Sir Edwin Alliott Verdon Roe was just A.V. Roe when he arrived at Squantum with his triplane design from England. Roe first gained the inspiration to design airplanes after serving with the merchant navy and witnessing the seemingly never-ending soaring flight of albatrosses. In 1907, Roe won a *Daily Mail* contest for flying model design, the prize being £75. (MAHS.)

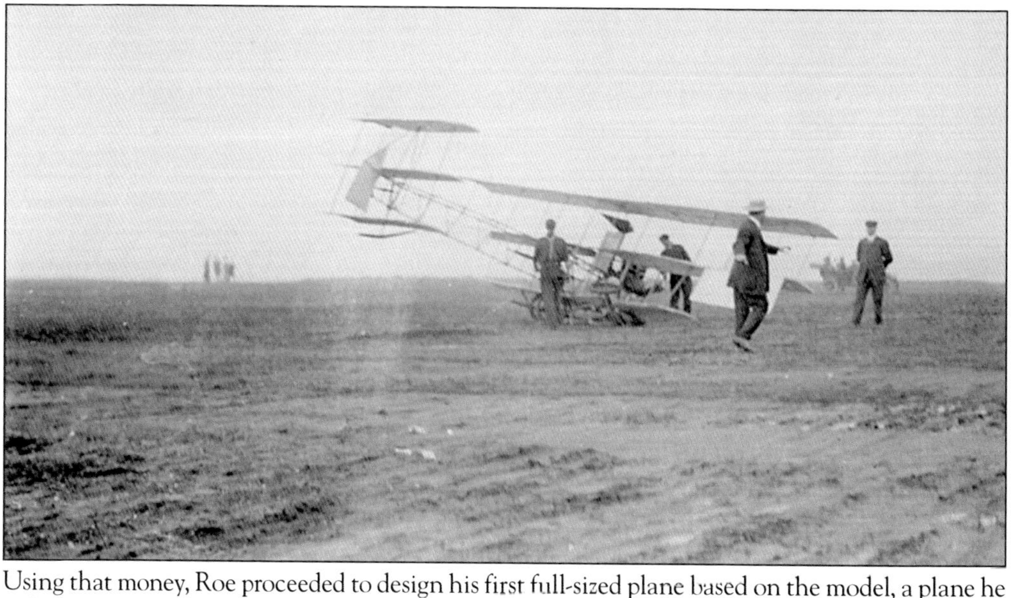

Using that money, Roe proceeded to design his first full-sized plane based on the model, a plane he called the Roe I Biplane. He tested it in 1907 and 1908, finally making a successful flight on June 8, 1908. On January 1, 1910, he founded A.V. Roe Aircraft Company with his brother Humphrey and soon built his triplane, which met with an accident at Squantum. (MAHS.)

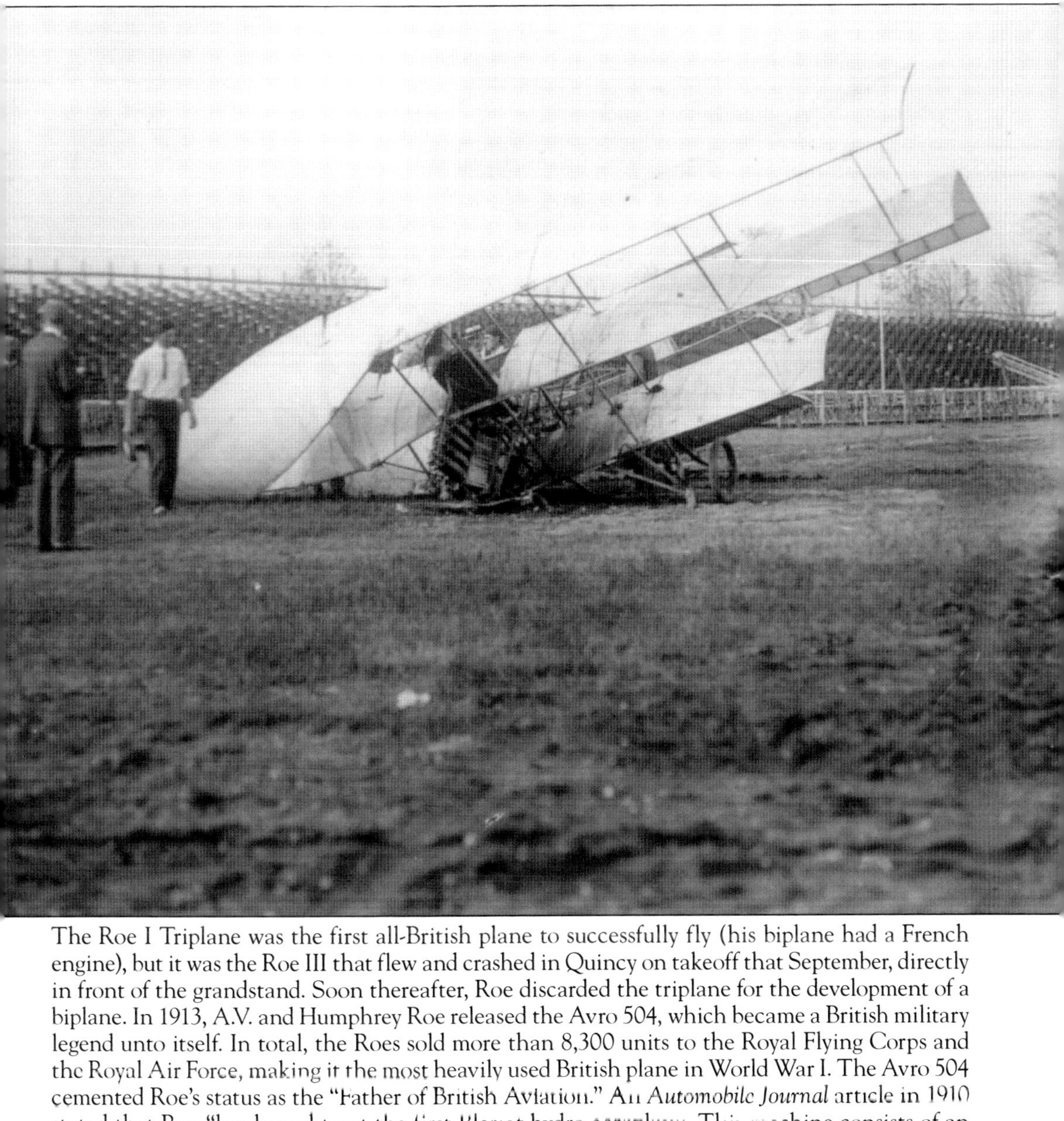

The Roe I Triplane was the first all-British plane to successfully fly (his biplane had a French engine), but it was the Roe III that flew and crashed in Quincy on takeoff that September, directly in front of the grandstand. Soon thereafter, Roe discarded the triplane for the development of a biplane. In 1913, A.V. and Humphrey Roe released the Avro 504, which became a British military legend unto itself. In total, the Roes sold more than 8,300 units to the Royal Flying Corps and the Royal Air Force, making it the most heavily used British plane in World War I. The Avro 504 cemented Roe's status as the "Father of British Aviation." An *Automobile Journal* article in 1910 stated that Roe "has brought out the first Bleriot hydro aeroplane. This machine consists of an Avro biplane to which has been fitted a pair of floats under the forward end of the fuselage and two smaller floats near the tips of the lower main planes." A.V. Roe was knighted in 1927. (QHS.)

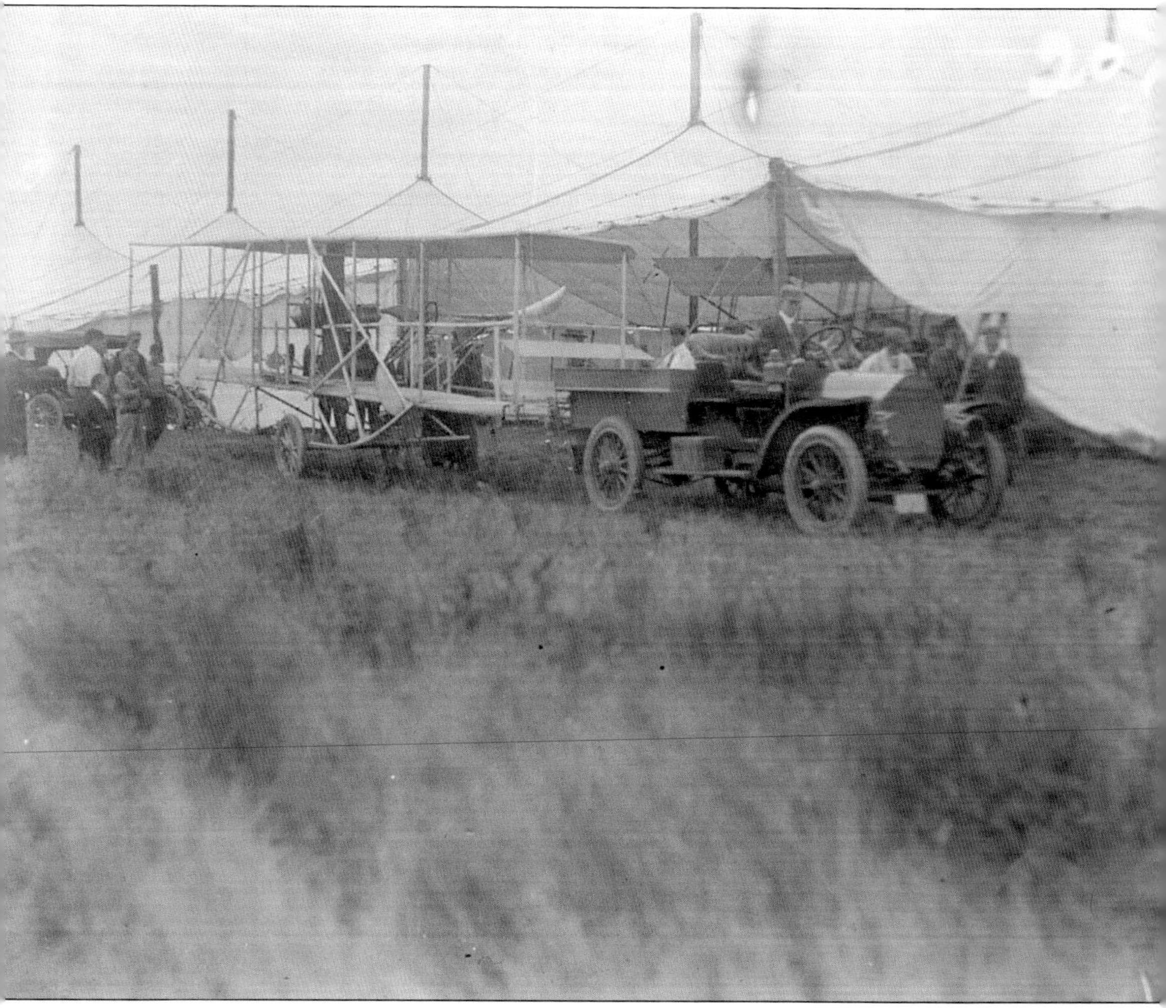

Among those finding it essential to be at the first Harvard Aero Meet was Wilbur Wright. Wright knew that not only would major European competitors be present, but so, too, would Glenn Curtiss, destined to be a major supplier of aircraft for the US Navy and reputed as the "Father of Naval Aviation," and Starling Burgess, who would sign a license agreement with the Wrights to manufacture their Model F at his Marblehead factory. Due to the huge number of military and government attendees, it was almost mandatory that Wright, Curtiss, and Burgess be present and provide their latest models for exhibition. Curtiss and Burgess each flew their own aircraft, but Wilbur Wright brought his flying team of Ralph Johnstone and Walter Brookings. Johnstone, a trick bicycle rider, took to the sky as a daredevil, unfortunately meeting his end just two months after the first Squantum meet, on November 17, 1910. Unable to fully pull out of a dive in Denver, Colorado, he became the first American to die while piloting an airplane that crashed. (QHS.)

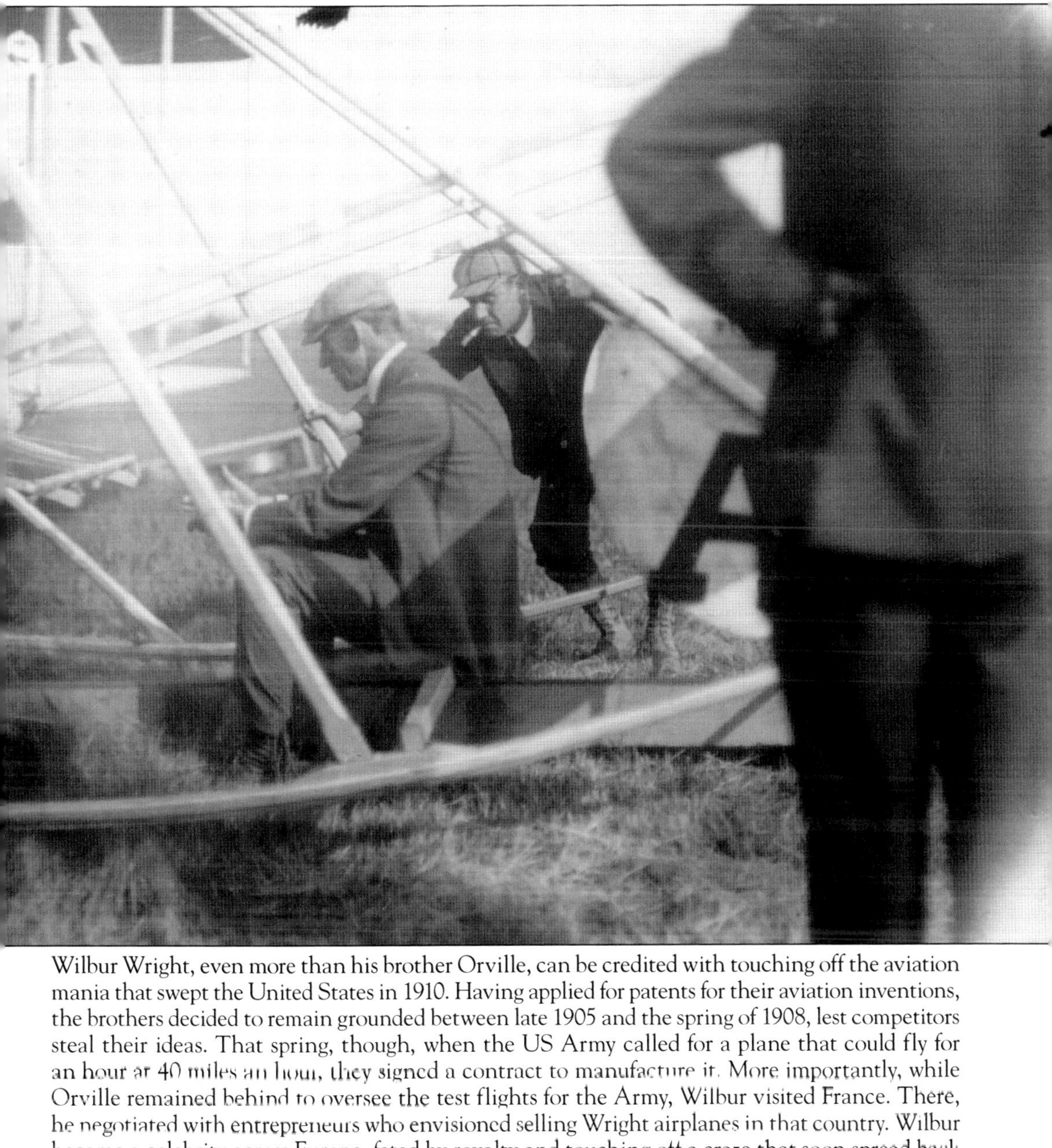

Wilbur Wright, even more than his brother Orville, can be credited with touching off the aviation mania that swept the United States in 1910. Having applied for patents for their aviation inventions, the brothers decided to remain grounded between late 1905 and the spring of 1908, lest competitors steal their ideas. That spring, though, when the US Army called for a plane that could fly for an hour at 40 miles an hour, they signed a contract to manufacture it. More importantly, while Orville remained behind to oversee the test flights for the Army, Wilbur visited France. There, he negotiated with entrepreneurs who envisioned selling Wright airplanes in that country. Wilbur became a celebrity across Europe, feted by royalty and touching off a craze that soon spread back across the Atlantic. Those folks who saw him at Squantum witnessed a shooting star; attacked by typhoid fever in May 1912, Wilbur died at the too-young age of 45. (QHS.)

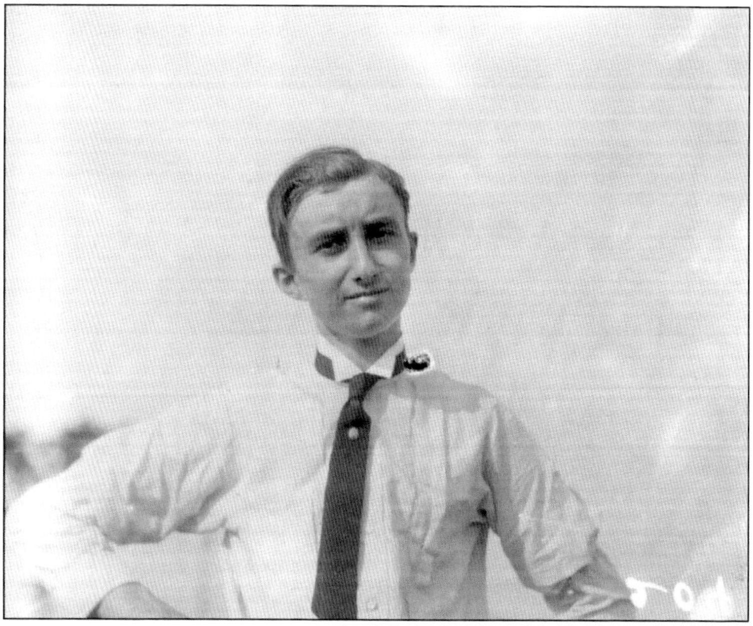

The Harvard-Boston Aero Meets featured a variety of individual contests ranging from speed-over-distance races; altitude; takeoff and landing distance; simulated bombing on a battleship outline on the field; rifle firing on ground targets; and even flights out to and around Boston Lighthouse and back. Among the entrants in the contests was Cromwell Dixon of Columbus, Ohio. (QHS.)

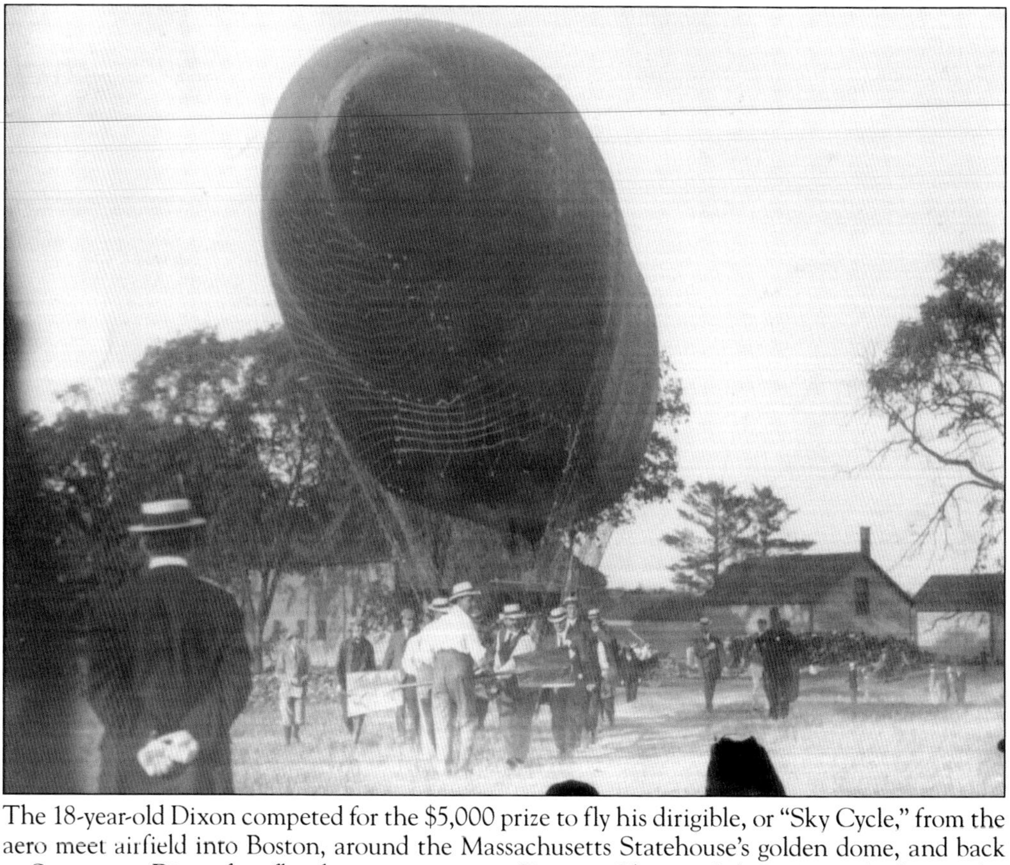

The 18-year-old Dixon competed for the $5,000 prize to fly his dirigible, or "Sky Cycle," from the aero meet airfield into Boston, around the Massachusetts Statehouse's golden dome, and back to Squantum. Dixon first flew his invention over Dayton, Ohio, on July 9, 1909, which was also his 17th birthday. (QHS.)

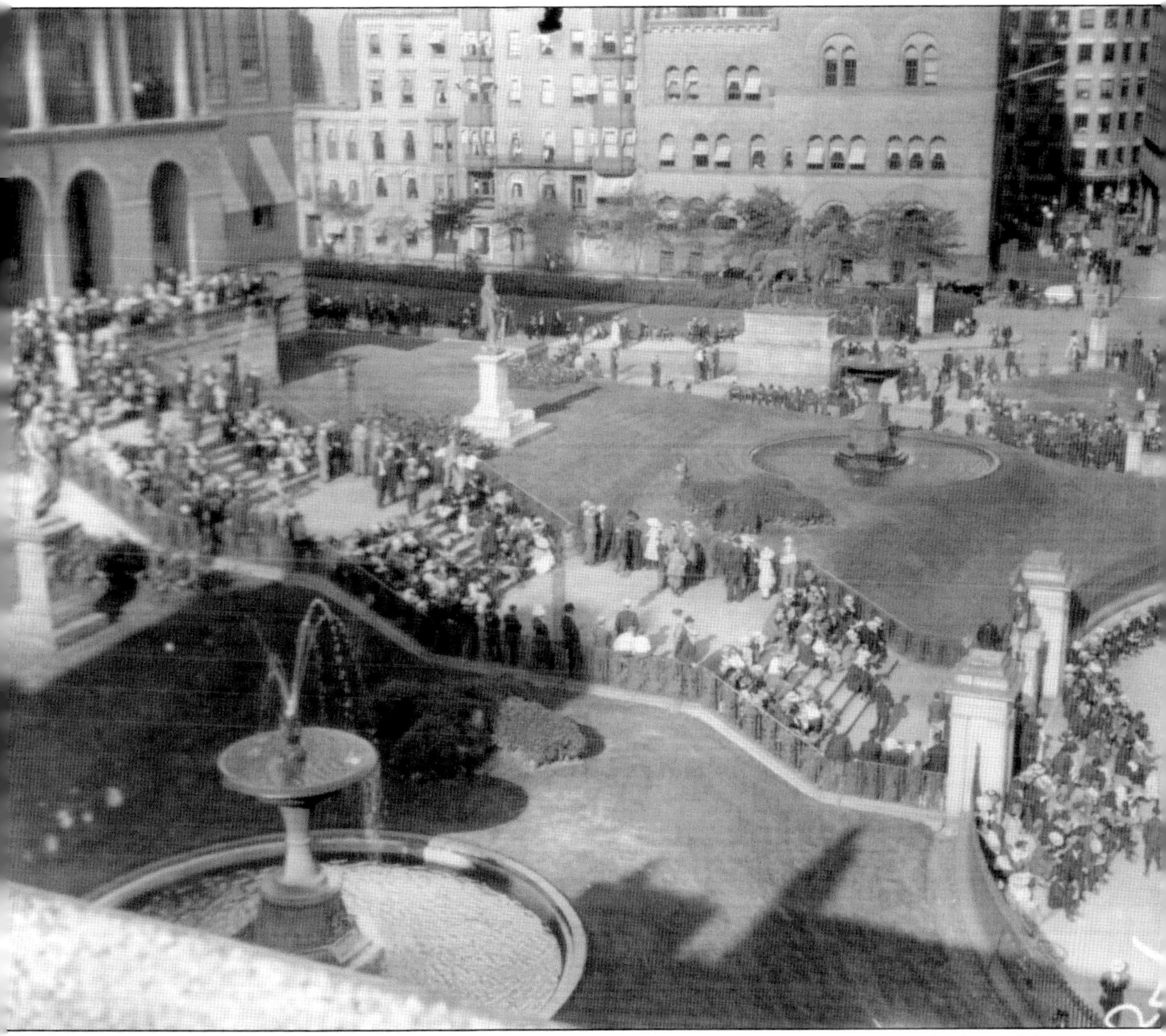

Dixon set out in grand style and proceeded to Boston. Unfortunately, he mistook the golden dome of the Christian Scientist Mother Church for the statehouse and returned to Squantum as many onlookers waited on the steps of the statehouse. He did not collect the prize money. Dixon's dirigible featured an envelope hand-sewn by his mother, who traveled with him to air meets around the country. Tragedy almost struck for Dixon at Squantum when the engine on his dirigible failed and it dropped slowly toward the water. He cleared land by 10 feet. In 1911, Dixon switched to HTA craft and became the first person to fly across the Continental Divide on September 30. But, like many early aviators, his luck soon ran out; Dixon died two days later, on October 2, when his plane was caught in a downwind in Spokane, Washington, and crashed from 100 feet. Dixon survived another hour but passed away in a hospital at just 19 years old. (QHS.)

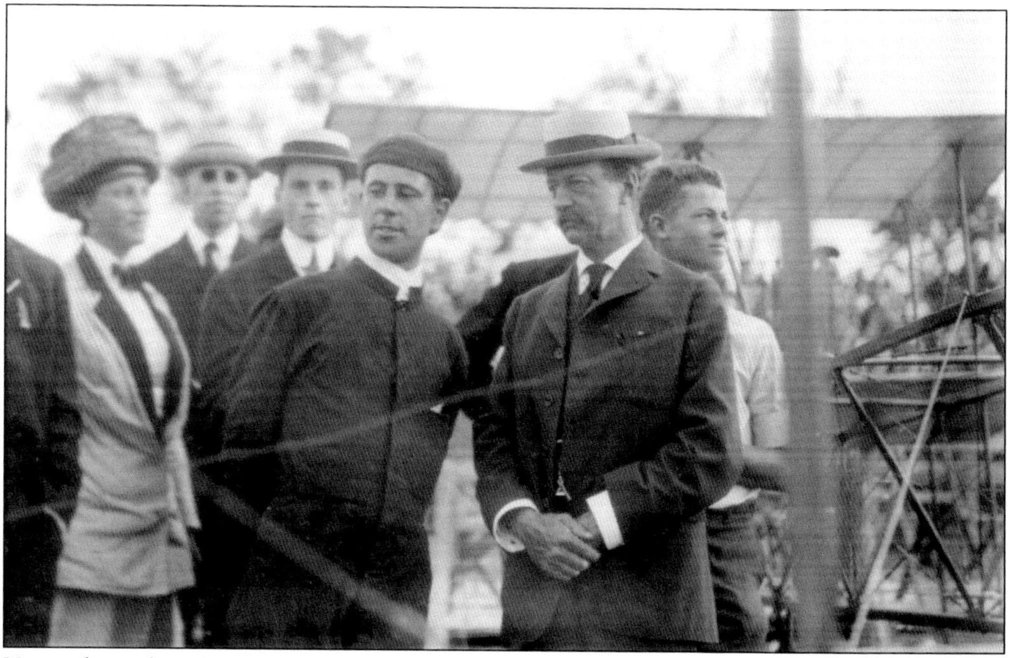

Harvard president A. Lawrence Lowell visited the airfield and instituted a rule that was somewhat controversial to the participants and attendees: no flying on Sundays. This caused some dismay and possibly had an effect on revenue. Even as one million were reported in attendance at the field, watching from surrounding high ground or in the streets, the goal of raising money to fund a department of aeronautics was not met, and the organizers lost about $22,000, mostly for field improvements. (QHS.)

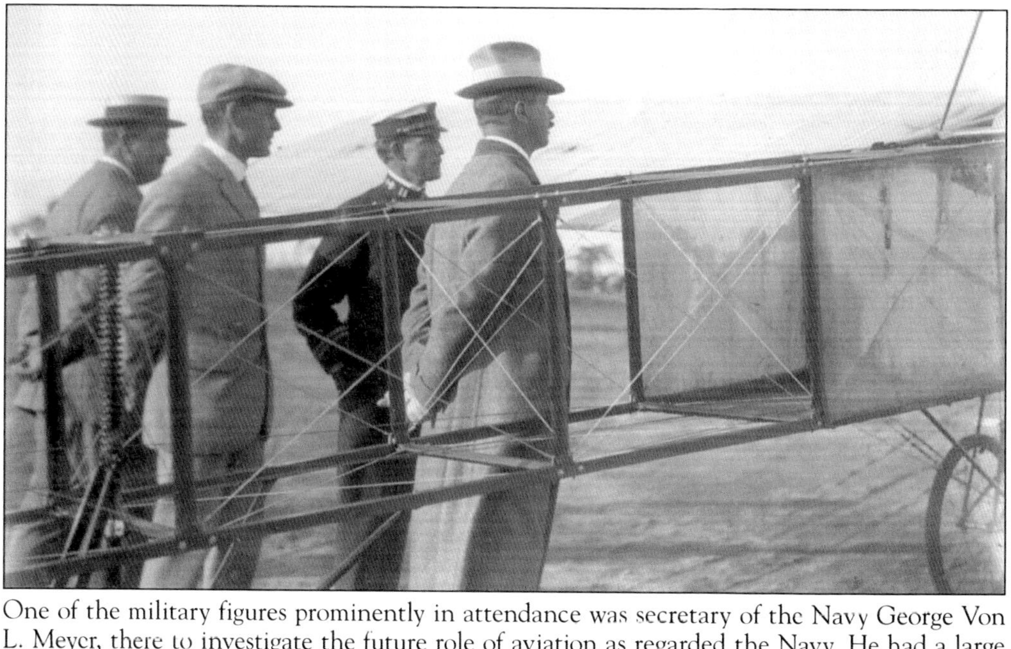

One of the military figures prominently in attendance was secretary of the Navy George Von L. Meyer, there to investigate the future role of aviation as regarded the Navy. He had a large contingent of officers with him, some assigned to the flying field, some assigned to patrol boats to observe flights over the water. (QHS.)

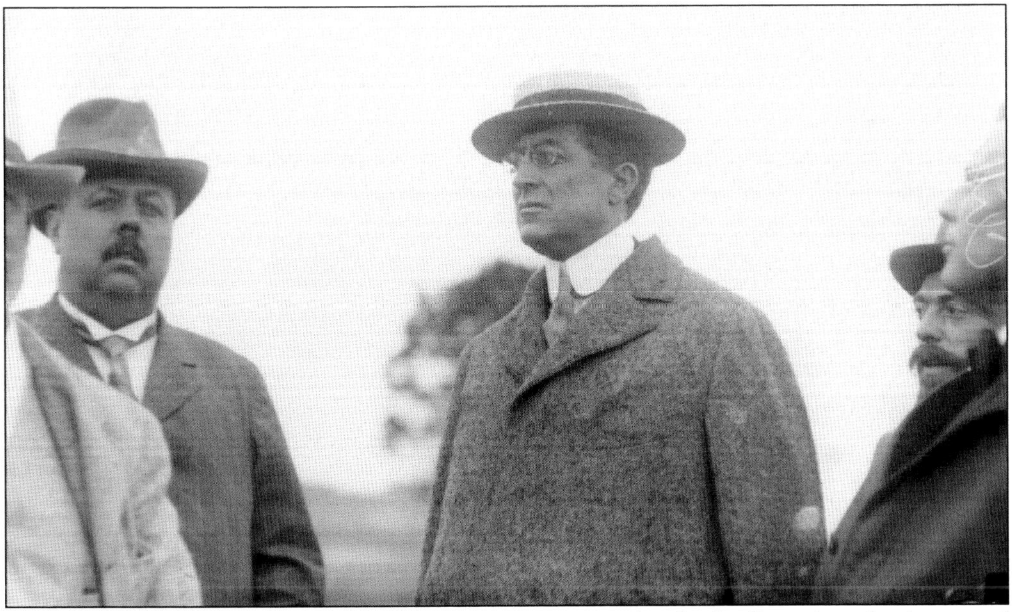

Clifford Burke Harmon was an old man for the flying game when he finally took to the sky around 44 years of age. In August 1910, a month before the first Squantum meet, Harmon became the first man to ever fly over Long Island Sound, a sound investment, indeed, for a man who made his fortune as a real estate developer in the New York area. (QHS.)

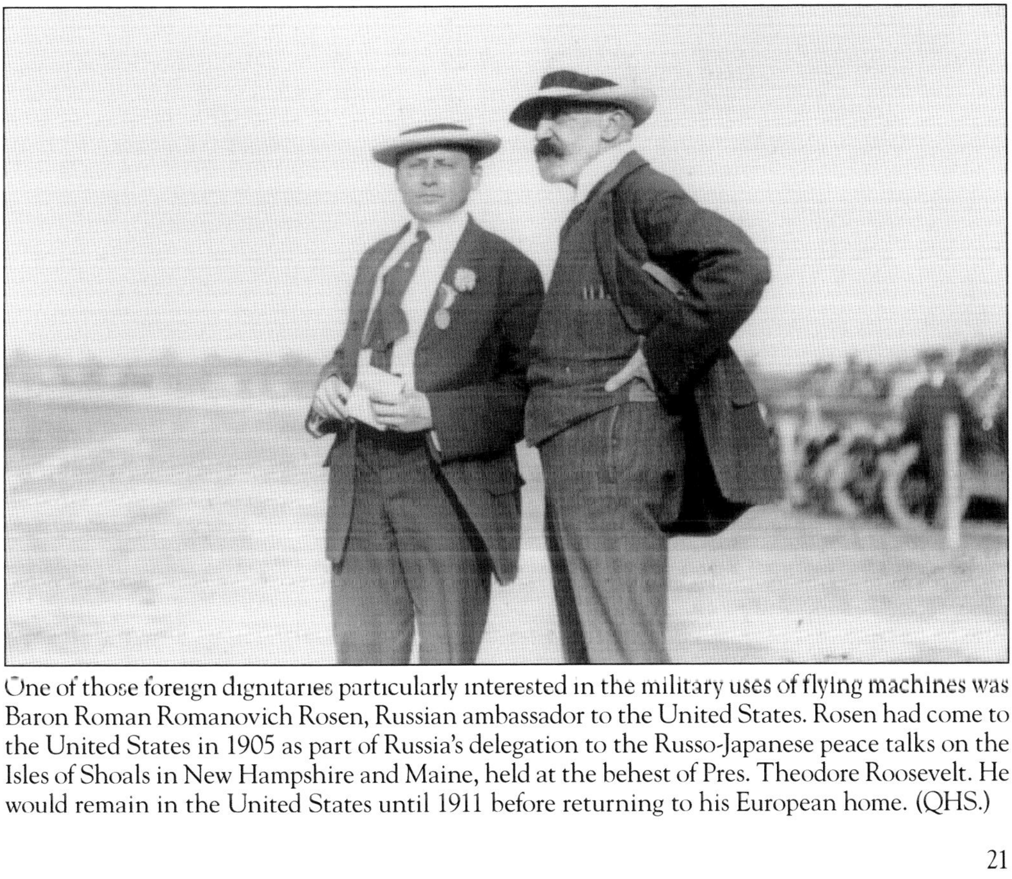

One of those foreign dignitaries particularly interested in the military uses of flying machines was Baron Roman Romanovich Rosen, Russian ambassador to the United States. Rosen had come to the United States in 1905 as part of Russia's delegation to the Russo-Japanese peace talks on the Isles of Shoals in New Hampshire and Maine, held at the behest of Pres. Theodore Roosevelt. He would remain in the United States until 1911 before returning to his European home. (QHS.)

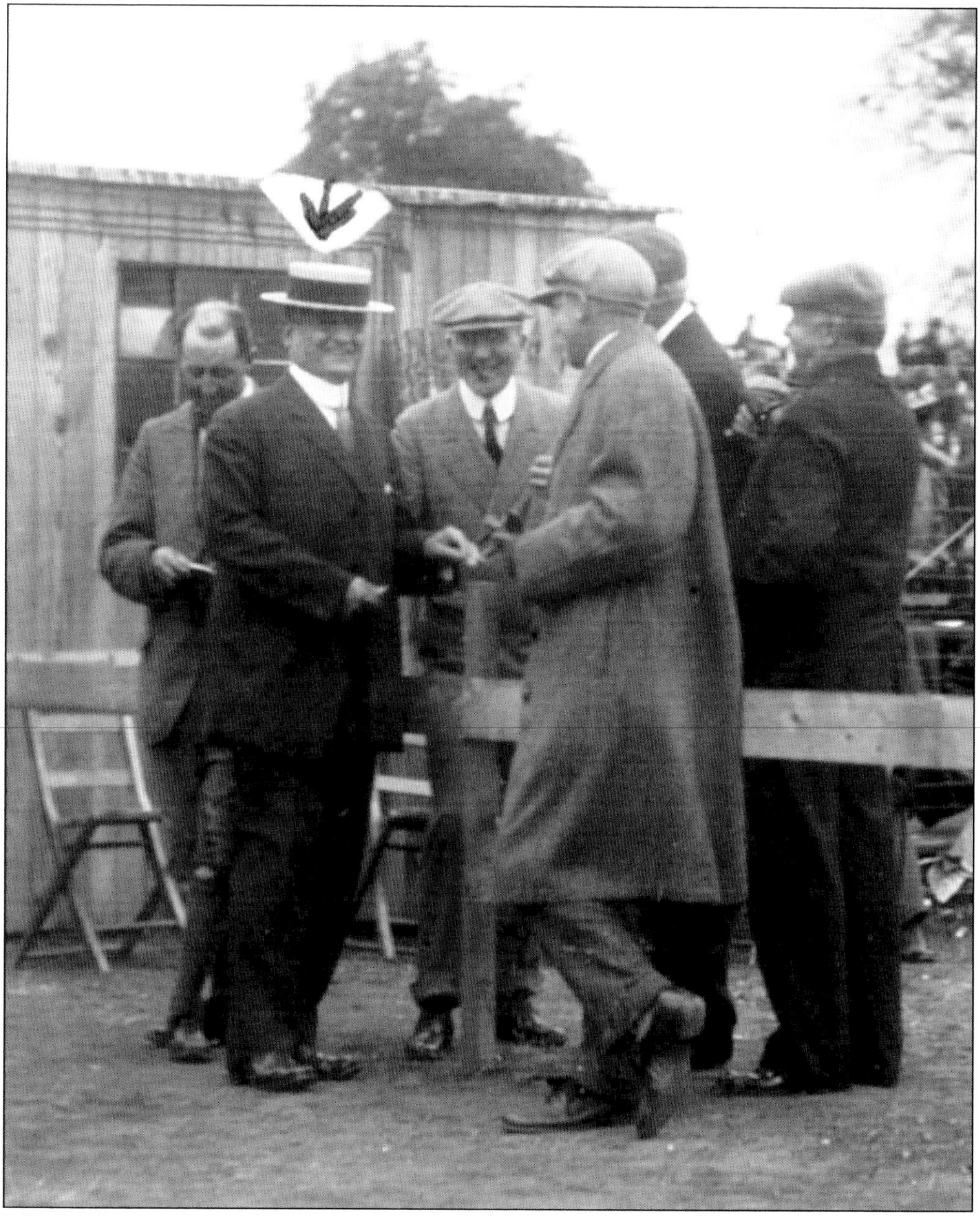

John F. "Honey Fitz" Fitzgerald knew how to find a crowd and knew what to do with it once he found it. The impish Irish politician glad-handed with the best of them, singing "Sweet Adeline" all the way to his first term as mayor of Boston from 1906 to 1908, and was on his way to his second term in 1910 when he arrived at the first Harvard Air Meet. Fitzgerald had more than political aspirations in mind when he met with flyers like Claude Grahame-White and other at the air meet. Under the slogan "A bigger, better and busier Boston," Fitzgerald launched his program for the commercial growth of the city during his first term. By 1912, the port was growing again. Although Fitzgerald never influenced aviation history, his grandson Lt. Joseph P. Kennedy Jr. died in a mid-air explosion in a BQ-8 (a converted B-24 Liberator) over France during World War II. (QHS.)

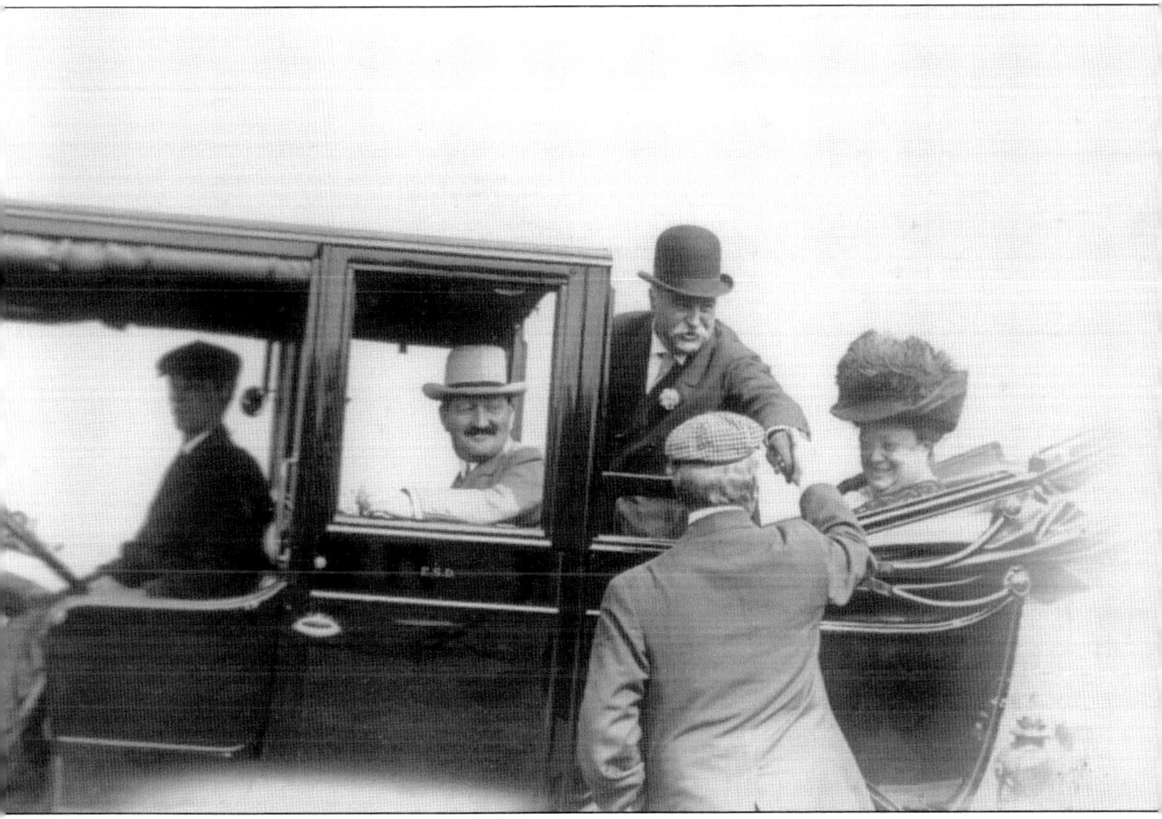

Eben S. Draper of Hopedale found numerous interests colliding at the first Harvard Air Meet. At the time governor of Massachusetts, Draper held a long history in manufacturing. A second-generation principal in the family business of manufacturing cotton machinery, Draper incorporated the Draper Company in 1887 (later the Draper Corporation), which became a major supplier of products to the Department of War. Draper, who studied at the Massachusetts Institute of Technology, joined the Massachusetts 1st Corps of Cadets in 1898, serving before and during the Spanish-American War. By the time of the Harvard Air Meet, the governor was on unsteady ground politically, mired in the fracturing of the state's Republican Party. An opponent to organized labor as union fever swept the country, Draper made decisions unpopular to the working man that led to his ousting from office in 1911. He lived just three more years, dying in Greenville, South Carolina, in 1914. (QHS.)

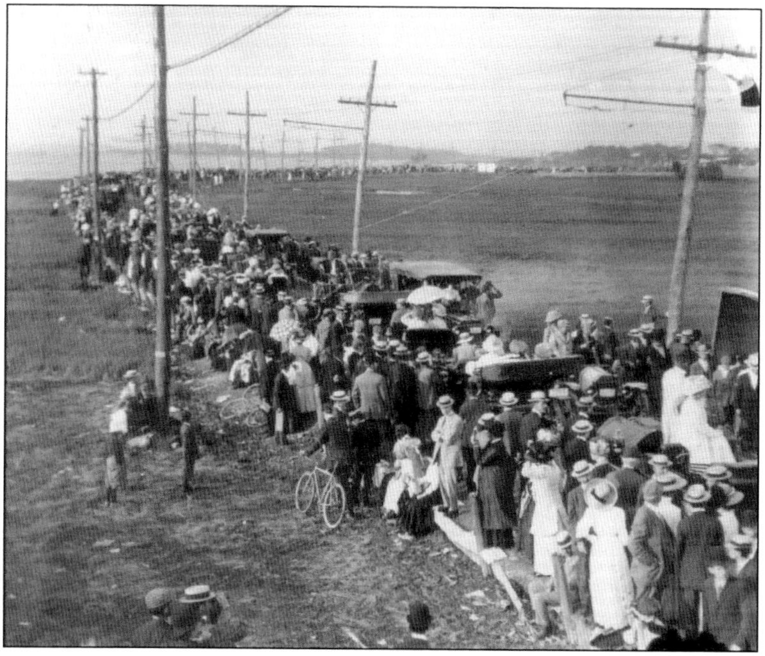

As far as the eye could see at Squantum, spectators rimmed the field. A sea of straw boaters along the shore indicated that it was summertime, as the fashion simply wouldn't do in winter. Spectators had gathered by any means possible, which included both Daisy on her bicycle built for two and Lucille in her merry Oldsmobile. (QHS.)

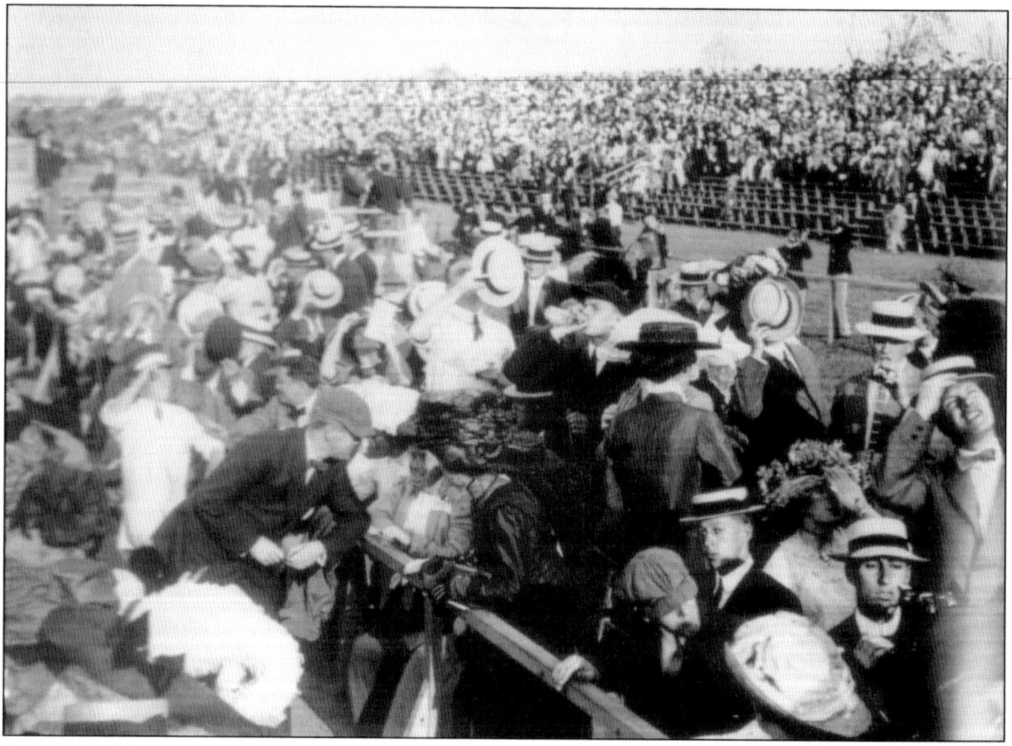

For most of the spectators, the attraction was pure spectacle. Humans had conquered the ground and sea. Bicycles, cars, and trains had shrunk distances once only passable by horse and carriage or on foot. Dreadnoughts roamed the seas, and submarines skulked below them. But man had only taken to the sky in 1903. A gathering of aviators plying their trade in 1912 was worthy of the excitement exhibited by this gathering. (QHS.)

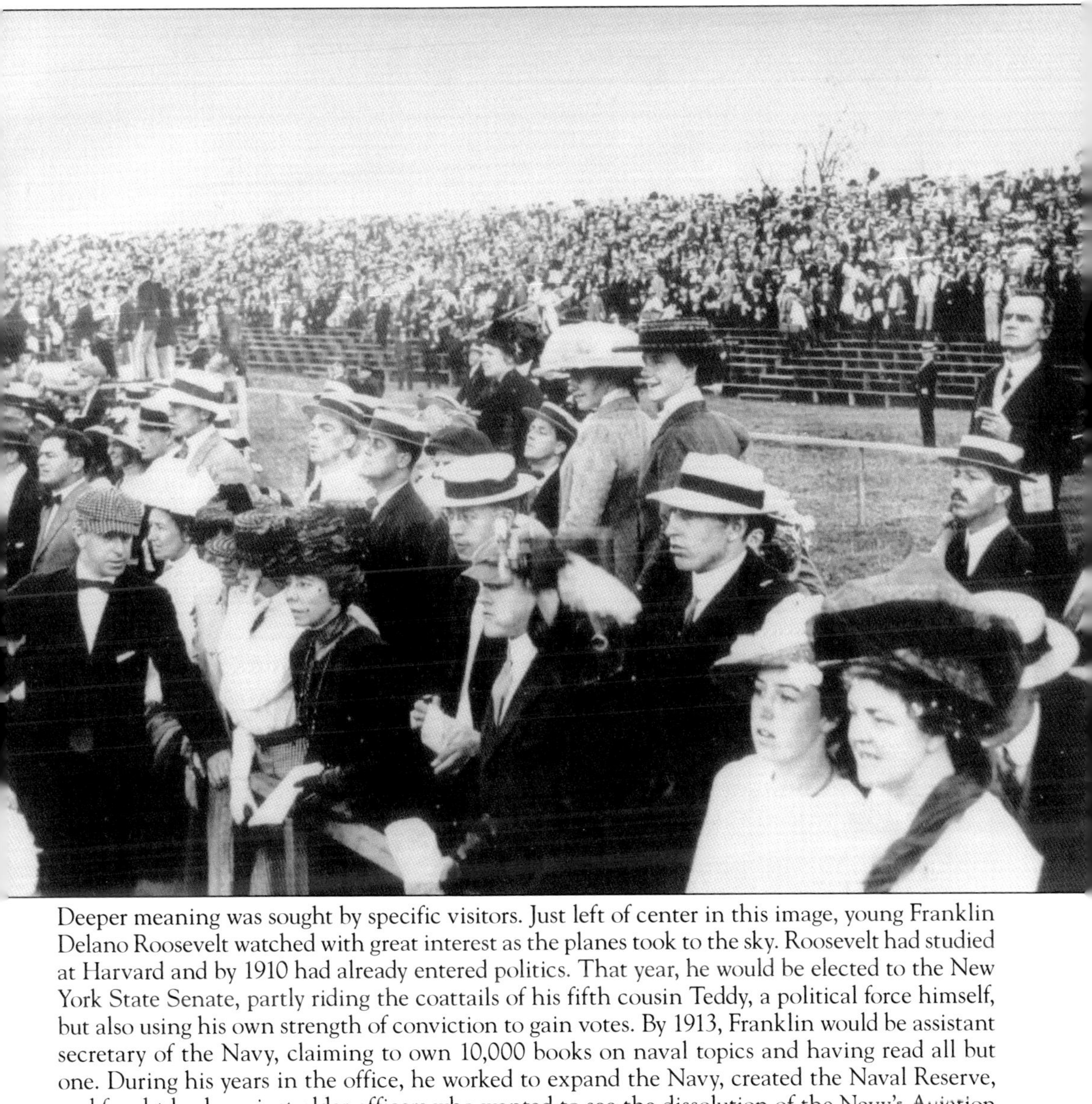

Deeper meaning was sought by specific visitors. Just left of center in this image, young Franklin Delano Roosevelt watched with great interest as the planes took to the sky. Roosevelt had studied at Harvard and by 1910 had already entered politics. That year, he would be elected to the New York State Senate, partly riding the coattails of his fifth cousin Teddy, a political force himself, but also using his own strength of conviction to gain votes. By 1913, Franklin would be assistant secretary of the Navy, claiming to own 10,000 books on naval topics and having read all but one. During his years in the office, he worked to expand the Navy, created the Naval Reserve, and fought back against older officers who wanted to see the dissolution of the Navy's Aviation Division during World War I. Though he did not at first believe in Army general Billy Mitchell's prediction that planes would one day sink ships with bombs, he believed in the power of military aviation, the basic tenets of which were displayed as he watched at Squantum in 1910. (QHS.)

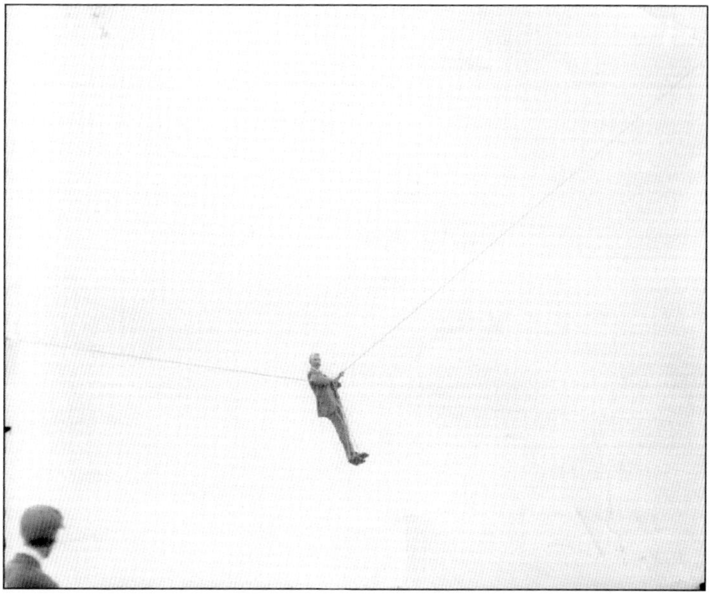

The Harvard Air Meet brought together daring folks of many kinds. Man-lifting kites originated in China as a method of execution for prisoners. By the 1890s, though, their use as both aerial observation platforms and sources of mirth were being explored. None other than Alexander Graham Bell developed one, and polar explorer Roald Amundsen tested another. Here, at Squantum, the kites were among the many forms of flight exhibited. (QHS.)

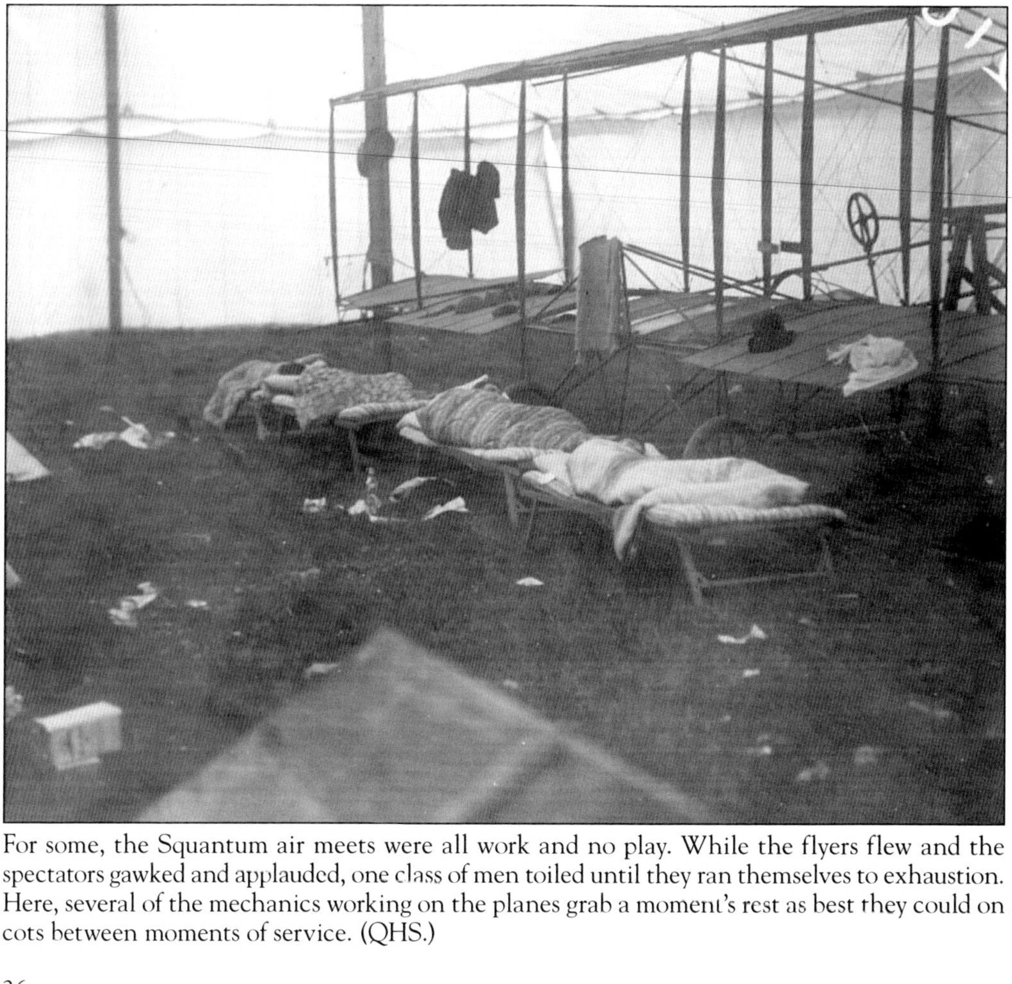

For some, the Squantum air meets were all work and no play. While the flyers flew and the spectators gawked and applauded, one class of men toiled until they ran themselves to exhaustion. Here, several of the mechanics working on the planes grab a moment's rest as best they could on cots between moments of service. (QHS.)

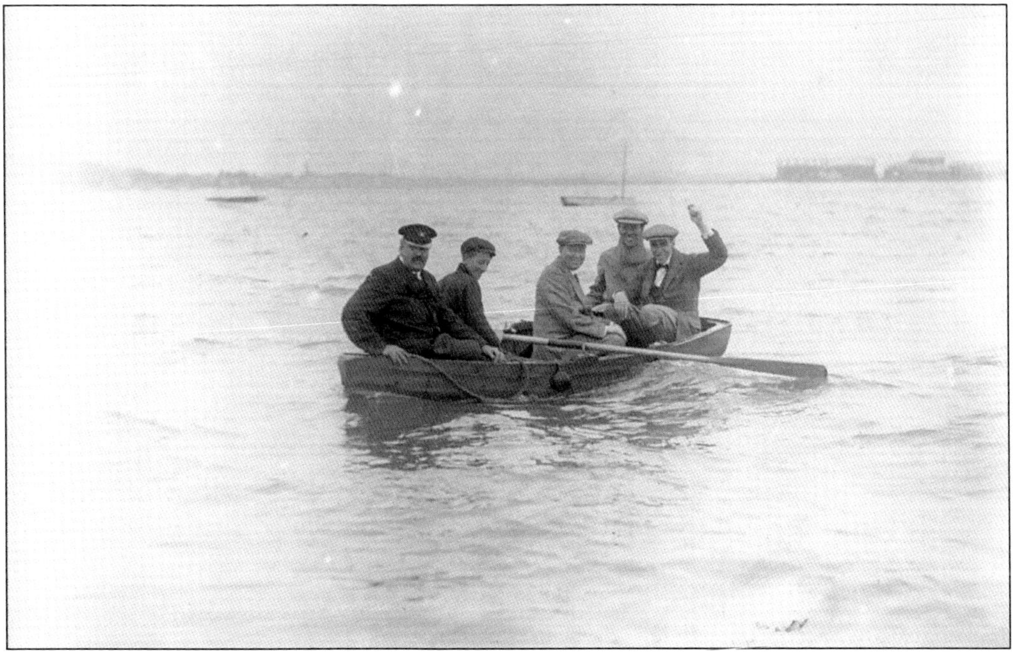

While the water might be more forgiving to a crashed plane than the solid ground, it of course came with its own hazards, especially for the pilots and their passengers. Luckily, Massachusetts was the birthplace of the American lifesaving system of mariners in distress at sea. A rescue boat was not hard to find and was on hand for any and all emergencies. (QHS.)

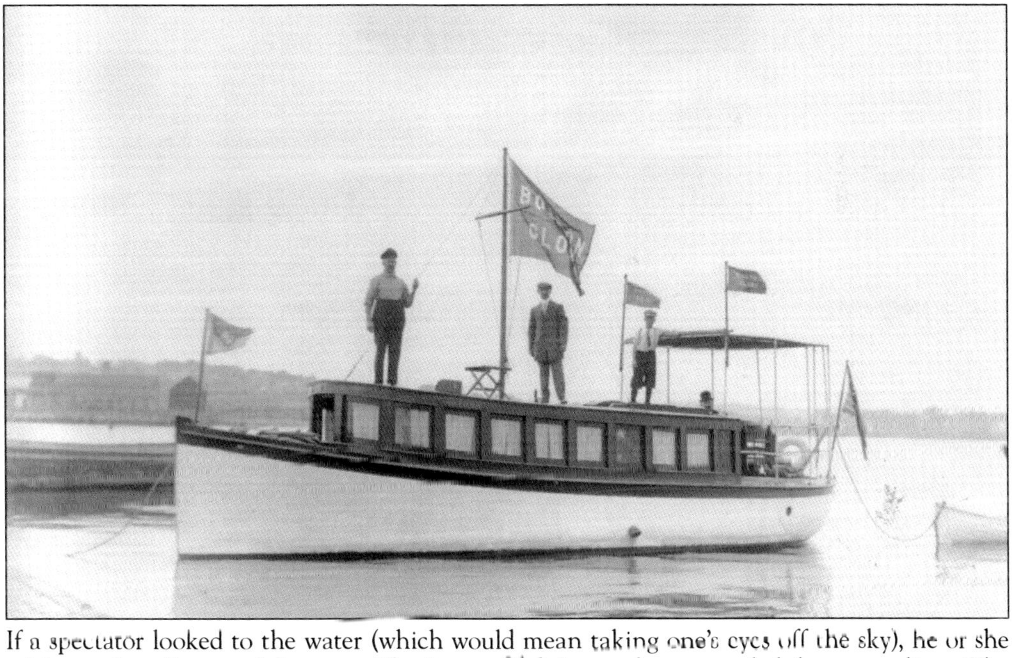

If a spectator looked to the water (which would mean taking one's eyes off the sky), he or she might catch a glimpse of the press. The *Boston Globe*'s press boat attended the proceedings. The paper, founded by Charles H. Taylor in 1872, served as one of numerous newspapers in the city but by 1890 had become a staple of the Bostonian's day, with an editorial staff emboldened by a plurality of Irish Catholics. (QHS.)

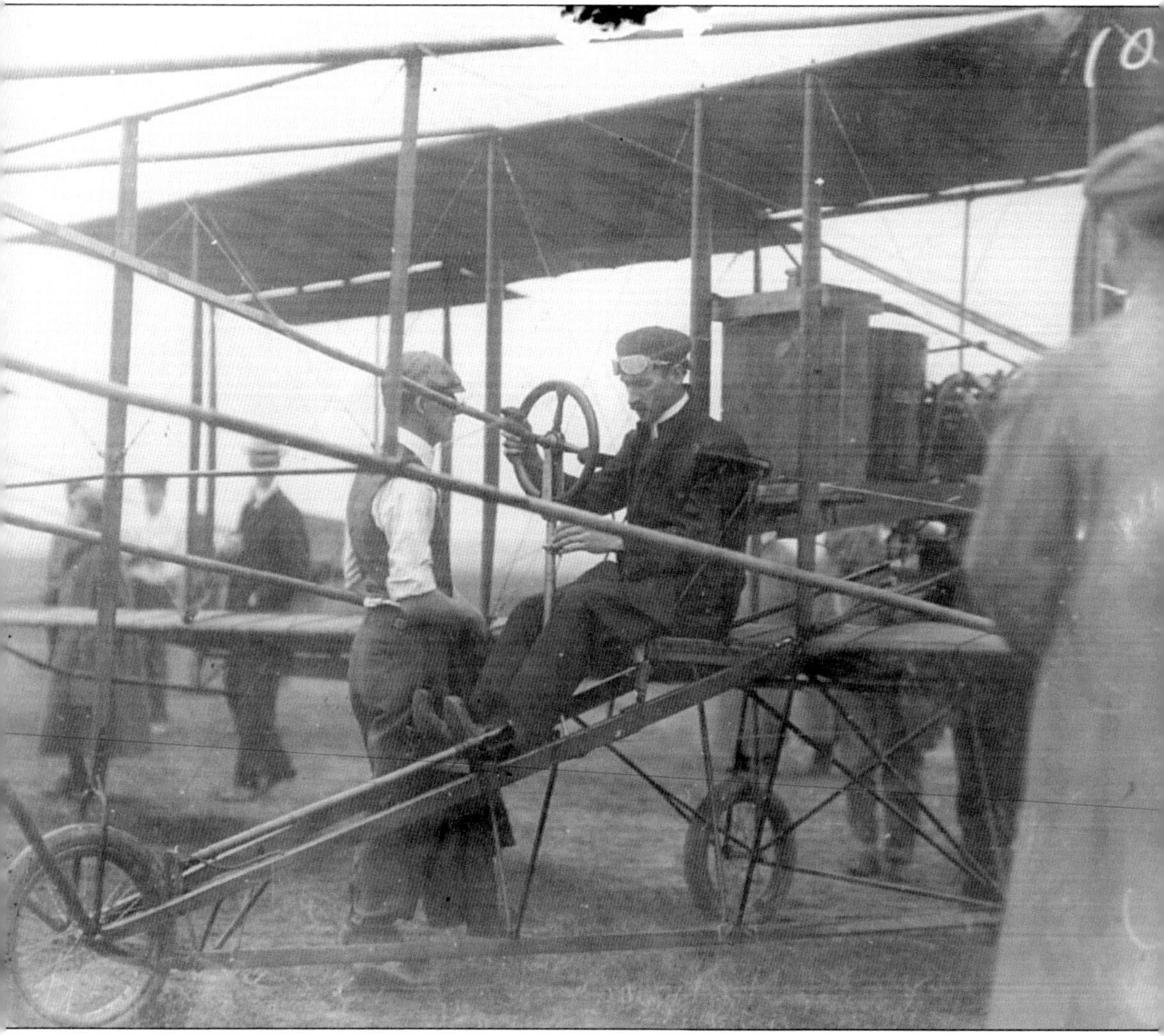

Like many early aviators, Glenn Curtiss made the transition from bicycles to planes. In Curtiss's case, the jump involved a period of motorcycle experimentation as well. The Wright brothers accomplished powered flight in 1903, but their progress was slowed, since they challenged every inventor from that time forward legally over patent infringement claims. Curtiss, though, kept at it and made many significant developments, winning the prestigious Gordon Bennet Trophy in France in 1909. In 1910, the Squantum meet for the first time brought together the three most famous people in American aviation: Wilbur Wright, Glenn Curtiss, and Massachusetts's own Starling Burgess. Curtiss would later be hailed as the "Father of Naval Aviation." His relationship with the Navy began later in 1910, when a pilot flying a Curtiss-designed plane launched and flew from a temporary deck built onto the cruiser USS *Birmingham*. That fall, Curtiss opened a winter training base in San Diego that would become known as the birthplace of naval aviation. (QHS.)

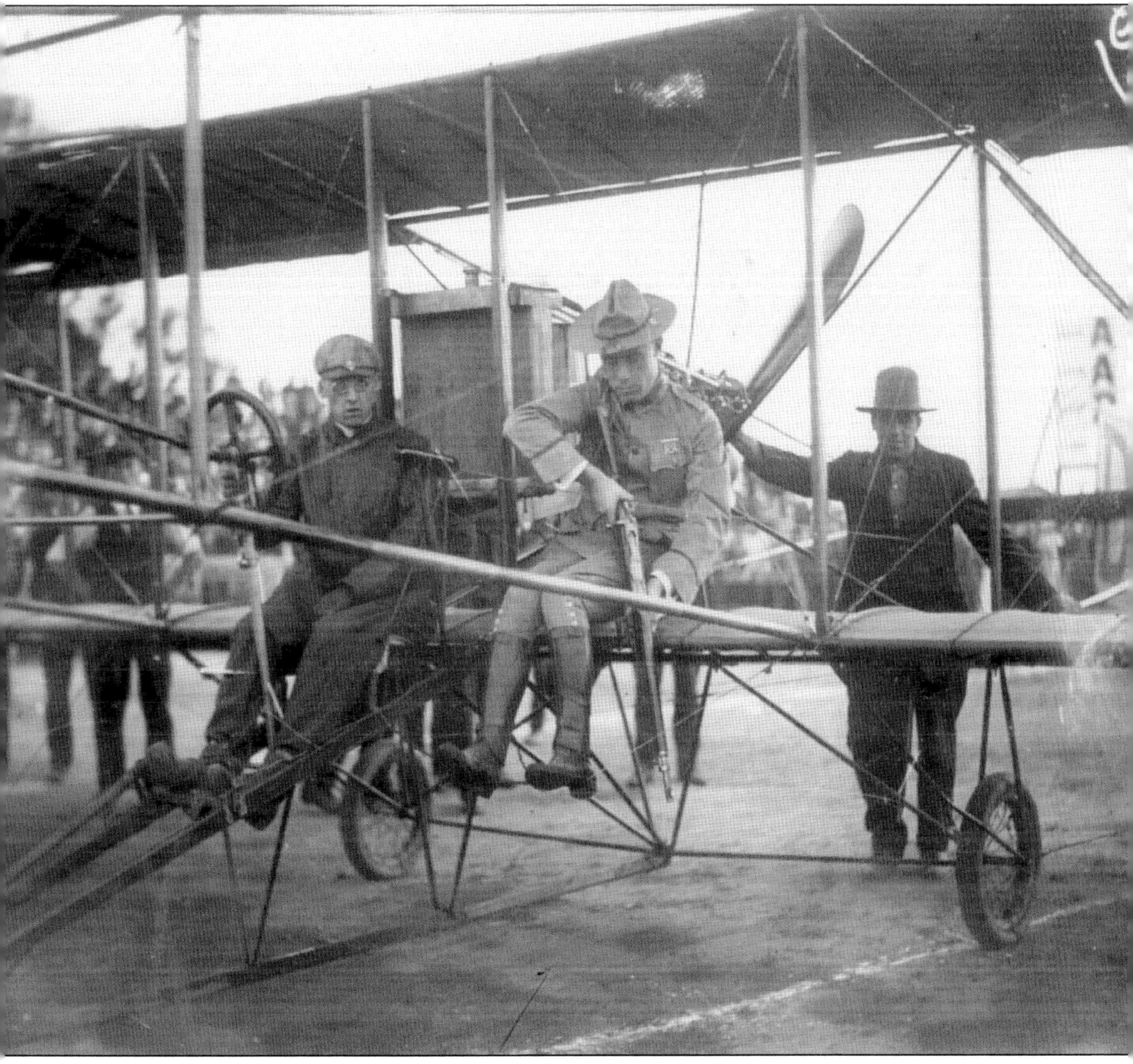

The question of whether or not the United States could use HTA aircraft for military purposes had already been pondered by the time of the first Harvard Air Meet. The US Army Signal Corps bought a Wright Model A in 1909. The plane became the first official military airplane in history. The Italians would be the first to use them in combat, in 1911 during their war with Turkey, and by Word War I, their use was ubiquitous. The question of how they should be used, though, was still being debated in 1910. Reconnaissance proved to be the simplest application, but bombing and even strafing would soon come. Here, Melrose's Charles Willard has the controls while Lt. Jacob Fickel demonstrates the art of firing from above. On August 20, 1910, Fickel, who would retire as a major general of the Army Air Force, fired the first ever rifle shot from an airplane, making him the United States' first aerial gunner. (QHS.)

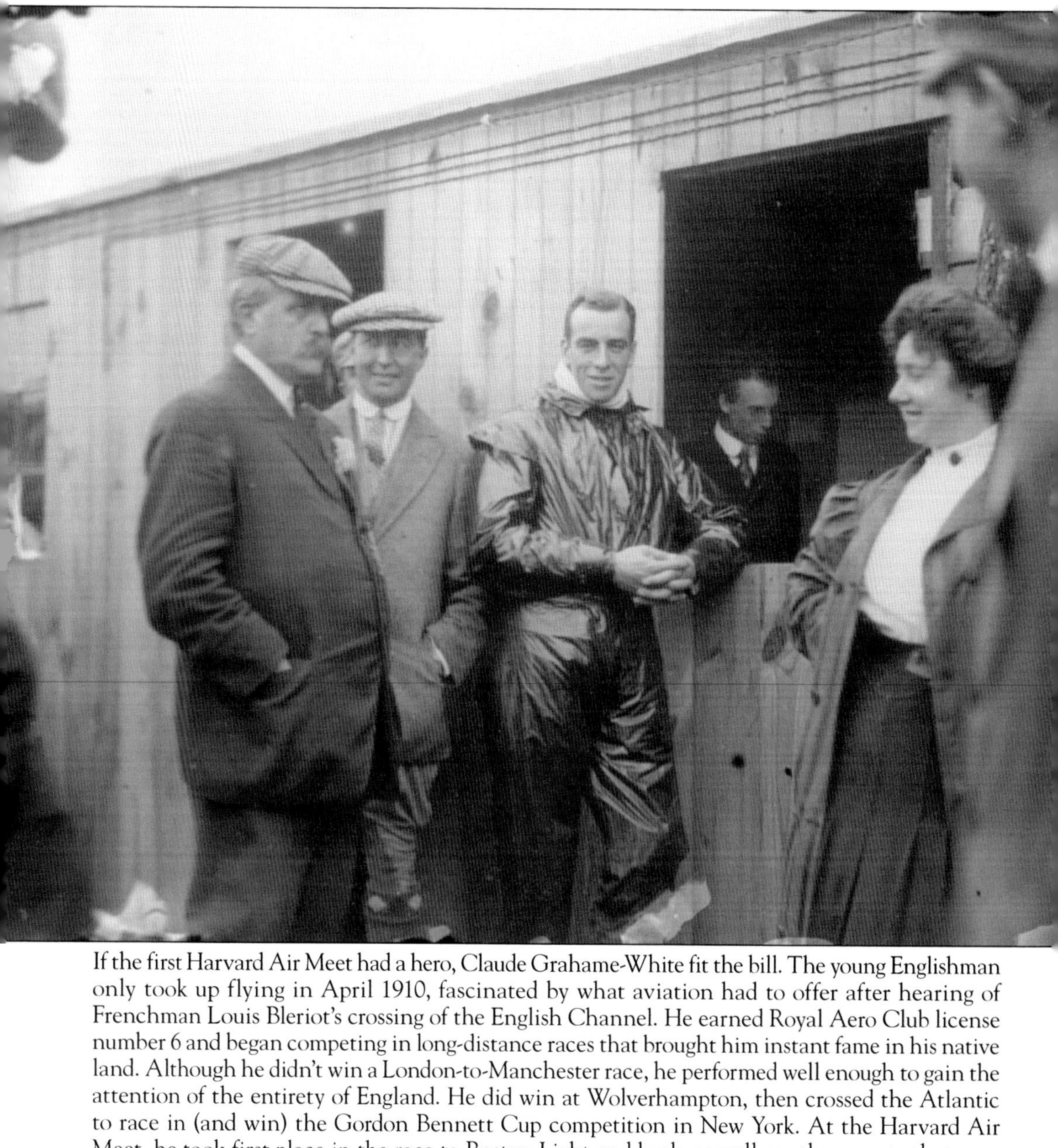

If the first Harvard Air Meet had a hero, Claude Grahame-White fit the bill. The young Englishman only took up flying in April 1910, fascinated by what aviation had to offer after hearing of Frenchman Louis Bleriot's crossing of the English Channel. He earned Royal Aero Club license number 6 and began competing in long-distance races that brought him instant fame in his native land. Although he didn't win a London-to-Manchester race, he performed well enough to gain the attention of the entirety of England. He did win at Wolverhampton, then crossed the Atlantic to race in (and win) the Gordon Bennett Cup competition in New York. At the Harvard Air Meet, he took first place in the race to Boston Light and back, as well as other events that gave him the overall crown for the meet. A few weeks after the Harvard meet, he landed a plane on Executive Avenue, near the White House in Washington, DC. Instead of being reprimanded for this act, he was applauded. (QHS.)

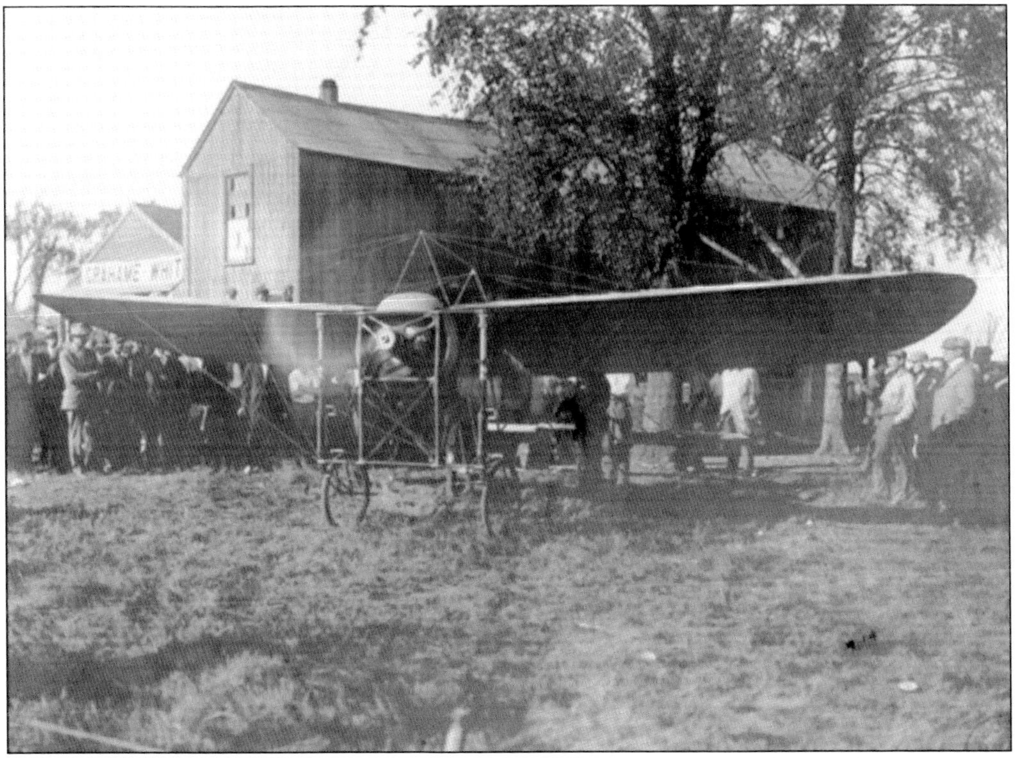

Claude Grahame-White's Bleriot monoplane was as famous as he was after the Harvard meet. Glenn Curtiss did not even attempt the flight to the lighthouse and back after Grahame-White set the time, knowing that his plane did not measure up. In all, at Squantum, Grahame-White won more than $29,000. Over the next three months, he won approximately $250,000 in exhibitions across the country. (MAHS.)

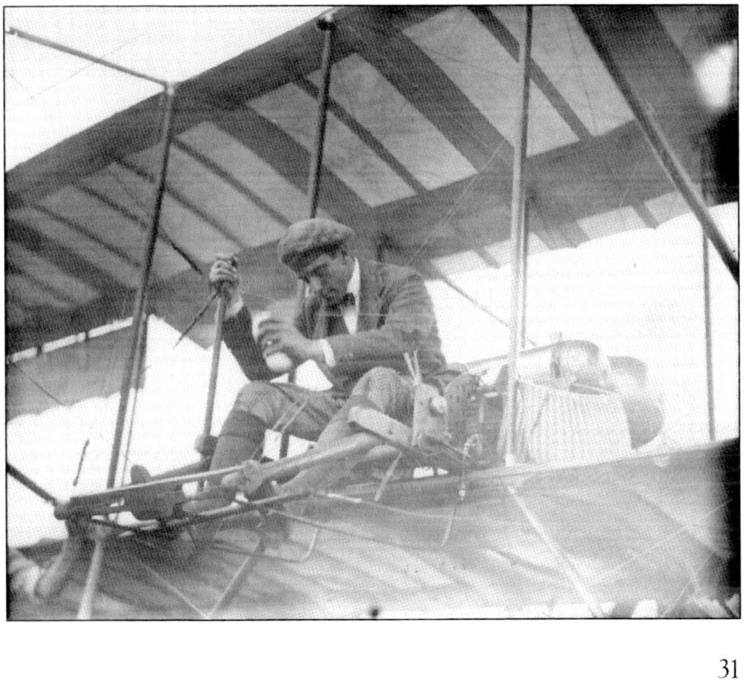

Grahame-White won several events at Squantum, including the bombing competition. The "bombs" were molded balls of plaster held in the wicker basket to his right in this picture. The target for the event was a wooden mock battleship. The first runs included drops from about 100 feet. Secretary of the Navy George Von L. Meyer challenged Grahame-White to hit the target from 1,800 feet, which he did. (QHS.)

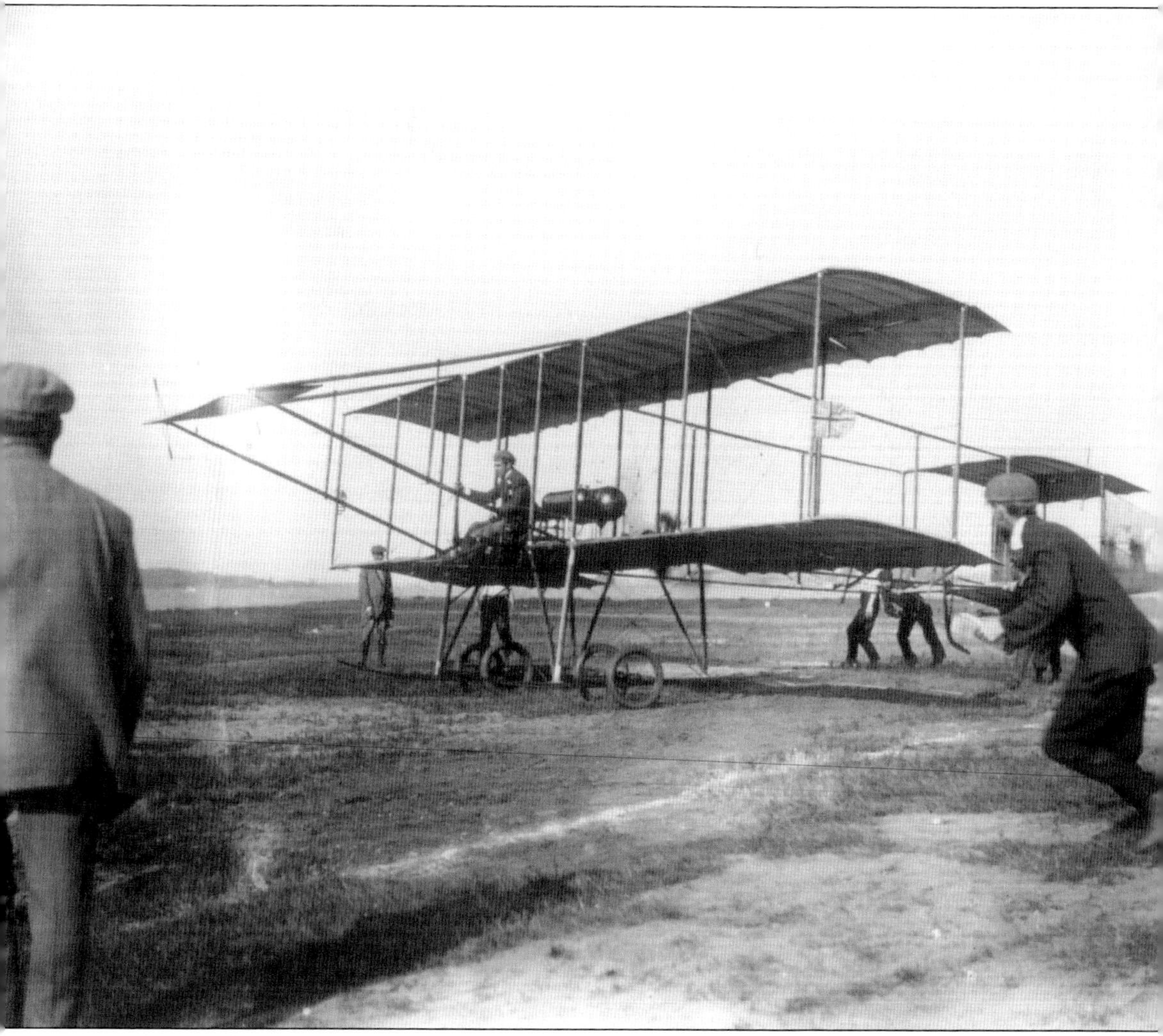

Grahame-White's prowess as an aviator allowed him the flexibility to fly different airplanes during the Harvard meet. Here, he takes off in his Farman III. Henry Farman based his first aircraft on one he purchased from Charles and Gabriel Voisin, the French owners of the first commercial aircraft factory in the world. Soon after buying the plane, Farman began making changes to its design, creating what became known as the Farman I or Voisin-Farman I. Further modifications led to the Farman III. That plane was made of ash with aluminum sockets holding the pieces together. Of the first 354 French pilots' licenses handed out, 81 were achieved using the Farman III. Only the Bleriot monoplane carried more would-be pilots (83) to their goal. Grahame-White flew his Farman in the London-to-Manchester race in April 1910, as did the pilot who beat him, Louis Paulhan. The Farman III even had military capabilities, used by the Greeks for reconnaissance in the Balkan Wars in 1912 and 1913. (QHS.)

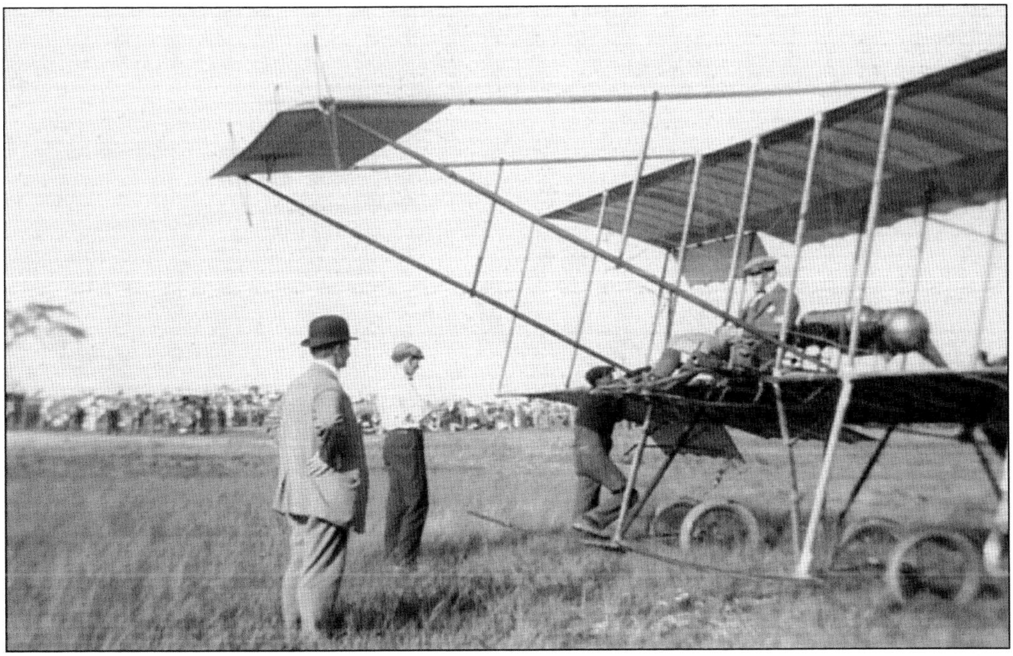

Among the modifications Farman made to the Voisin and Voisin-Farman I aircraft that led to the redesignation of the plane as the Farman III was the replacement of the original pairs of wheels with two long skids with pairs of wheels mounted with bungee cord and held in check by radius rods. (MAHS.)

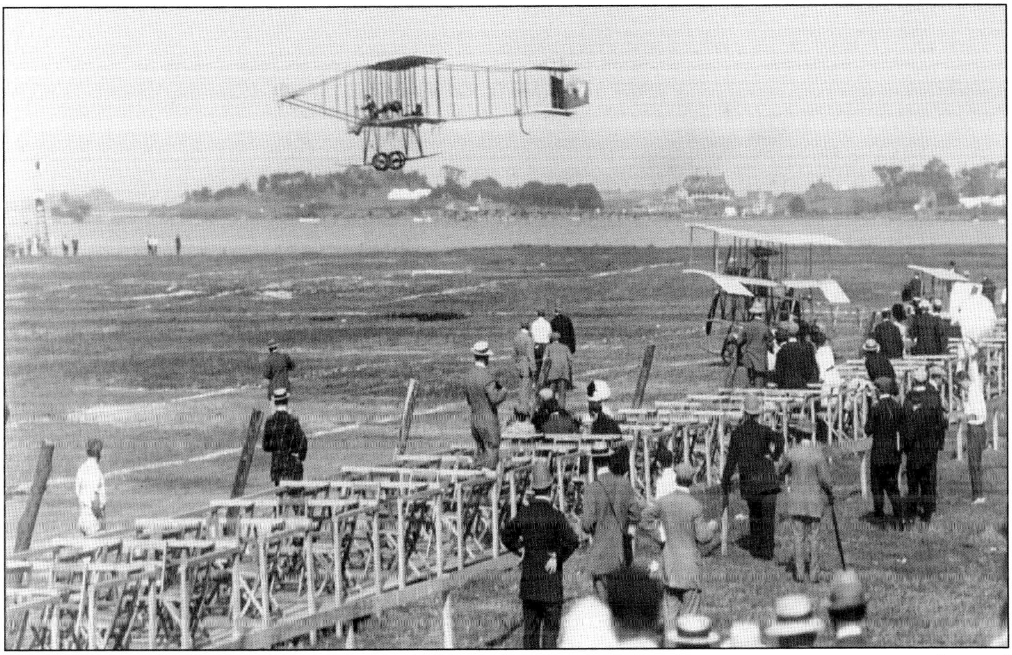

The Farman III in 1910 also featured a 50-horsepower Gnome Omega rotary engine in place of the 50-horsepower, inline four-cylinder water-cooled Vivinus originally installed. For comparison, the Cessna 172, the most successful aircraft in history, utilizes engines from 180 to 210 horsepower. Today, 50-horsepower engines are reserved for outboard motors for small motorboats. (MAHS.)

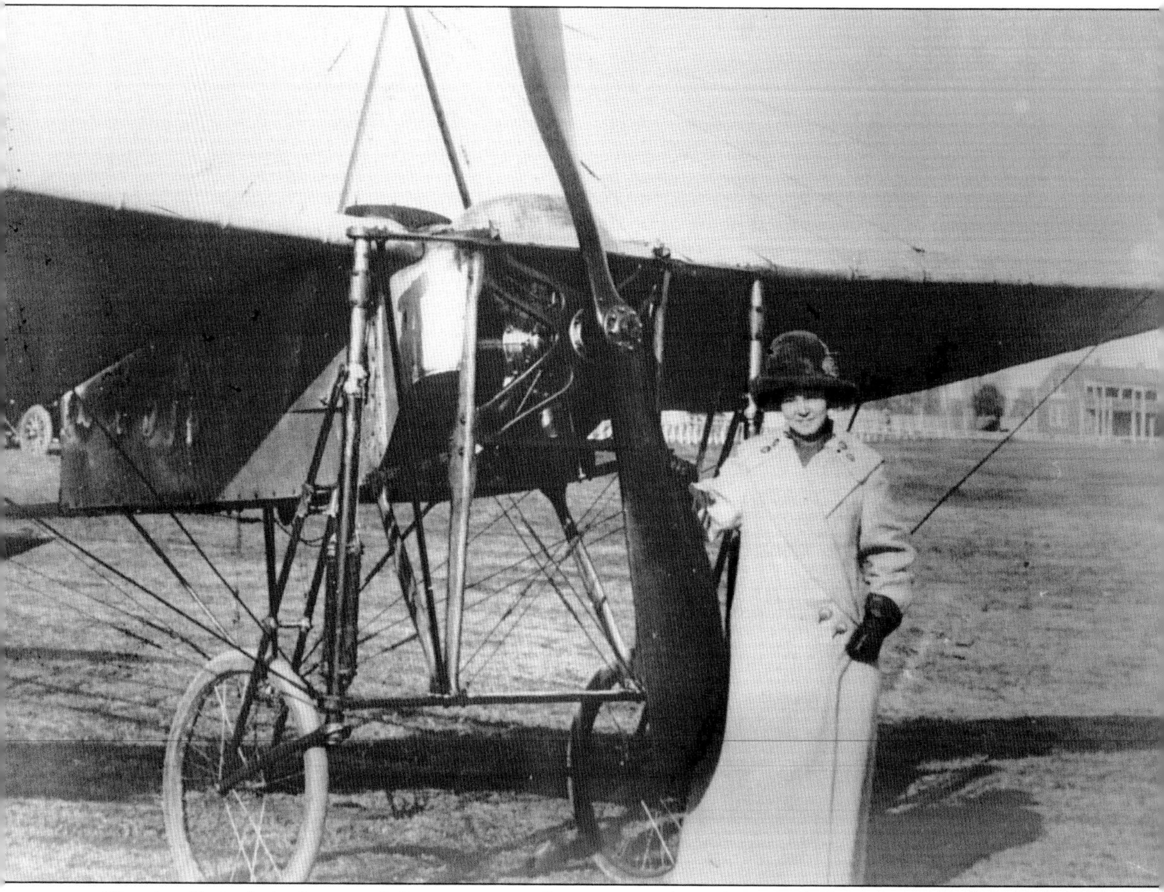

Harriet Quimby's dreams did not start with aviation, but they ended with it. By the time that Quimby registered to fly at the Third Annual Boston Aviation Meet at Squantum in 1912 (Harvard having dropped its affiliation with the event that year), Quimby had more than 250 articles under her belt for *Leslie's Illustrated Weekly*, where she was a theater critic for nine years. She also tried her hand at screenwriting, pulling together the scripts for seven silent films in 1911, all of which were directed by D.W. Griffith, a pioneer of modern filmmaking. In 1910, Quimby attended the Belmont Park International Aviation Tournament on Long Island and fell in love with aviation. On August 1, 1911, she became the first licensed female pilot in America. Wearing a vibrant purple flying suit, she caught the attention of Vin Fiz, the grape soda company, for whom she became the new spokesperson after the death of Calbraith Perry Rodgers. Rodgers was the first man to fly coast-to-coast in the continental United States. (QHS.)

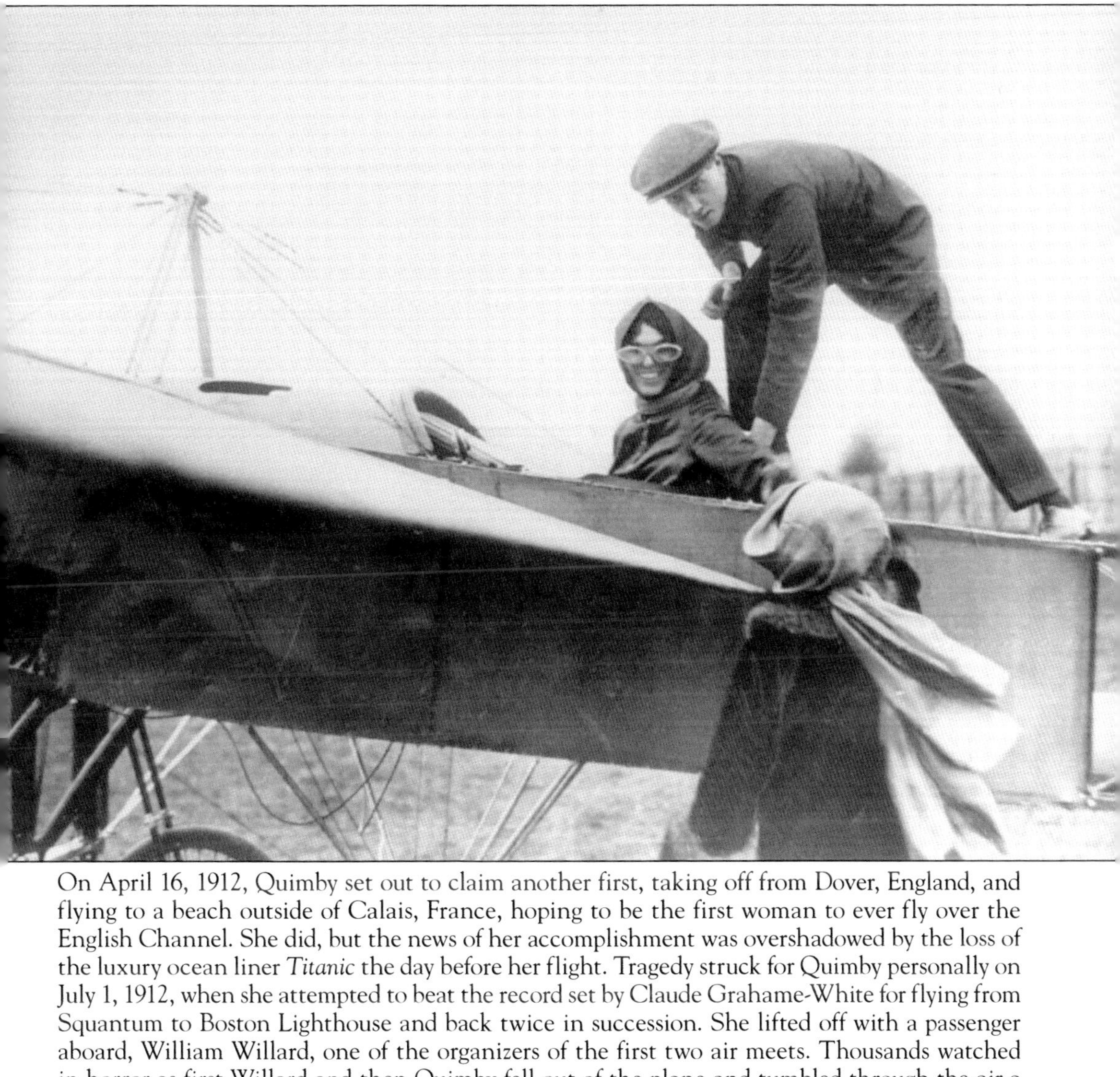

On April 16, 1912, Quimby set out to claim another first, taking off from Dover, England, and flying to a beach outside of Calais, France, hoping to be the first woman to ever fly over the English Channel. She did, but the news of her accomplishment was overshadowed by the loss of the luxury ocean liner *Titanic* the day before her flight. Tragedy struck for Quimby personally on July 1, 1912, when she attempted to beat the record set by Claude Grahame-White for flying from Squantum to Boston Lighthouse and back twice in succession. She lifted off with a passenger aboard, William Willard, one of the organizers of the first two air meets. Thousands watched in horror as first Willard and then Quimby fell out of the plane and tumbled through the air a thousand feet to their deaths. The catastrophe marked the end of the air meets at Squantum. Nonetheless, Quimby had made her mark as a pioneer in the field, opening opportunities for other women to take to the sky. (QHS.)

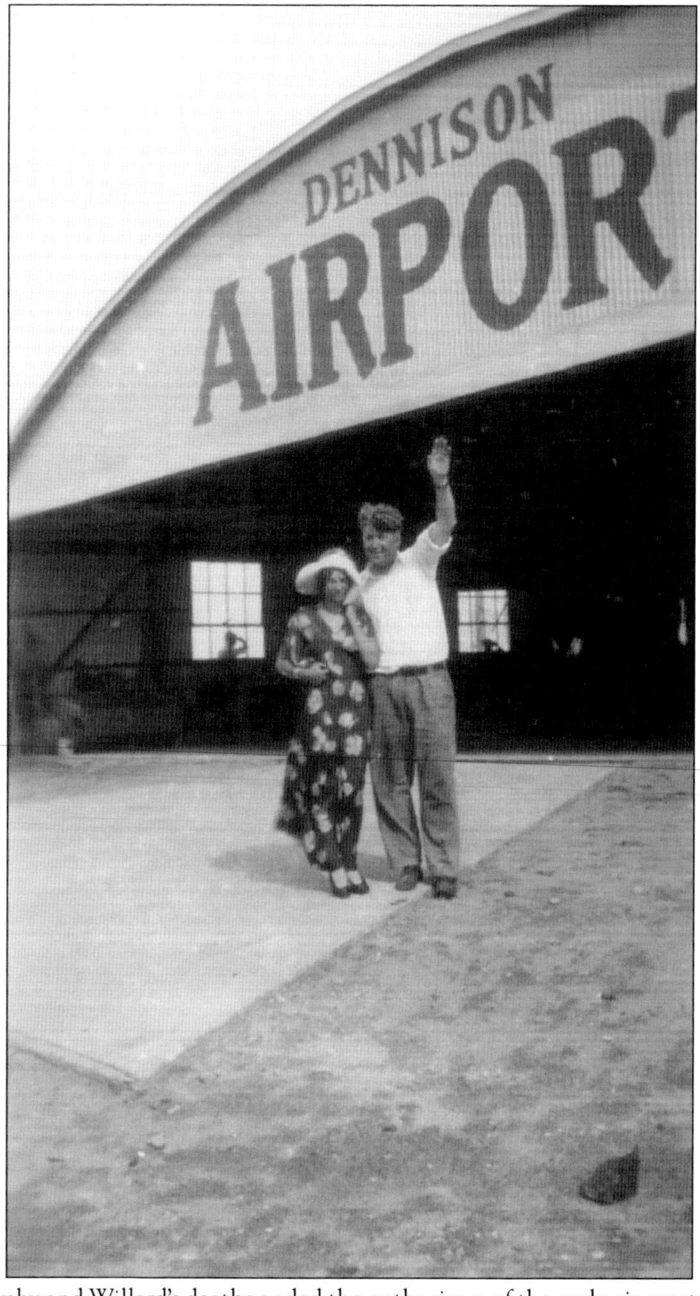

Although Quimby and Willard's deaths ended the enthusiasm of the early air meets at Squantum, civilian aviators would return to the area in 1927. The US government, which by then operated the Victory Destroyer Plant and a Naval Reserve airfield at Squantum, leased nine acres of land to Quincy architect Harold T. Dennison for use as an airfield. Dennison offered flight training classes, air shows, and scenic excursions over the coastline. In October 1923, Dennison had been in a training accident at the new Squantum base, ditching into the sea when his wing clipped a sandbar in Dorchester Bay. He and his training officer swam and crawled to shore, where they were taken to the hospital to recover. Dennison's own training classes stressed safety precautions while in the air. Here, Doris Shalit-Oberg and Paul Wilcox stand outside the Dennison hangar. (QHS.)

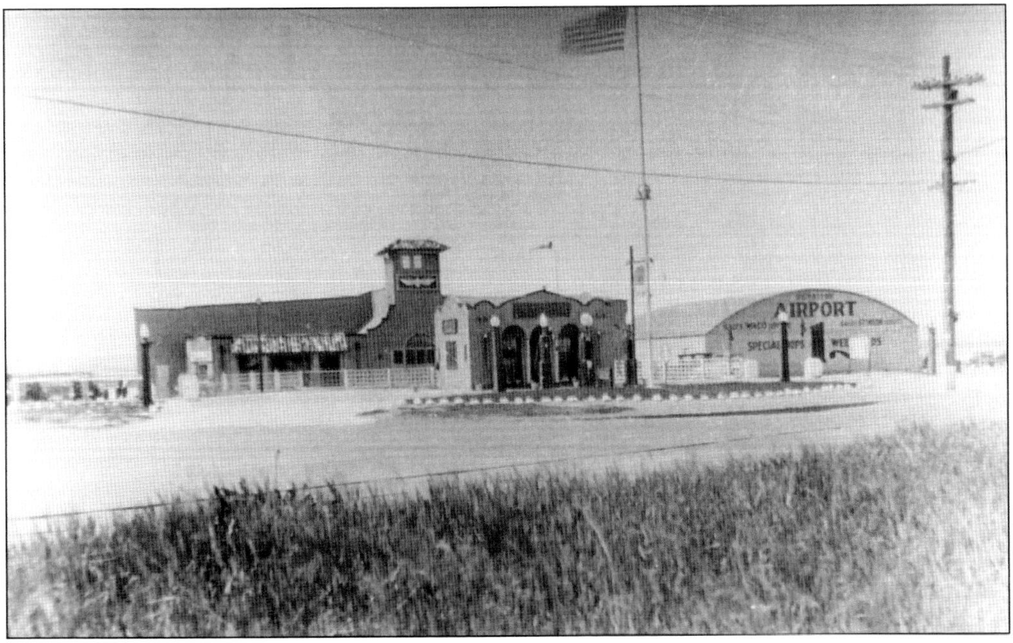

The airport sat near the intersection of today's East Squantum Street and Quincy Shore Drive and had star power from the start. One of the first investors, living in Medford at the time, Amelia Earhart helped finance its construction. She flew on the first flight out of the airport on September 3, 1927, and in 1932, when attempting the first transatlantic solo flight by a female, she flew over the airport and waggled her wings as a salute. (QHS.)

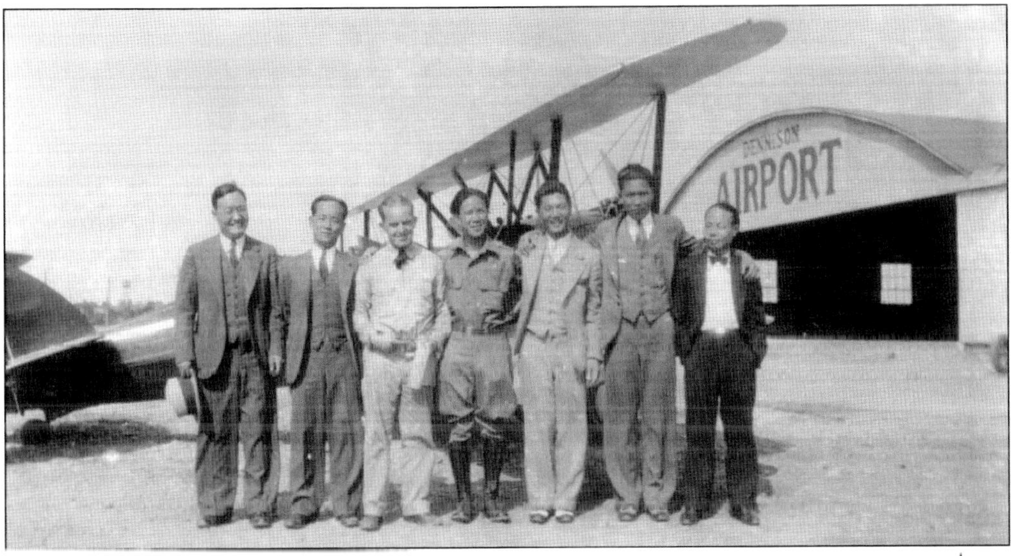

The appearance of this team of interested Japanese gentlemen at Dennison Airport was a novelty in 1930. Both the United States and Japan were in deep economic recessions. The Japanese people at the time were disappointed with their politicians and looking to their military for future guidance. With the congeniality shown here, it is obvious none of the photograph's subjects could foresee a time in the near future when the countries would become bitter enemies. (QHS.)

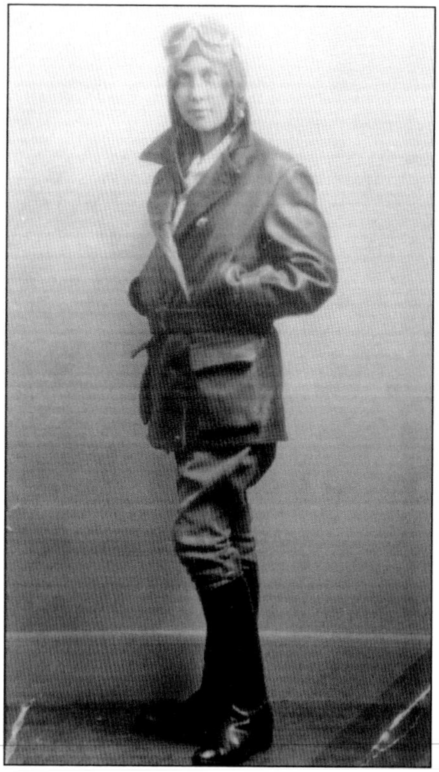

Doris Shalit-Oberg, at just five feet tall, became the youngest woman to fly solo out of Dennison Airport, which she did while she was still in high school in 1930. Shalit-Oberg became a Quincy legend and a president of the city's historical society. In 1991, at 78 years old, she accomplished a lifelong goal, skydiving from 10,000 feet. (QHS.)

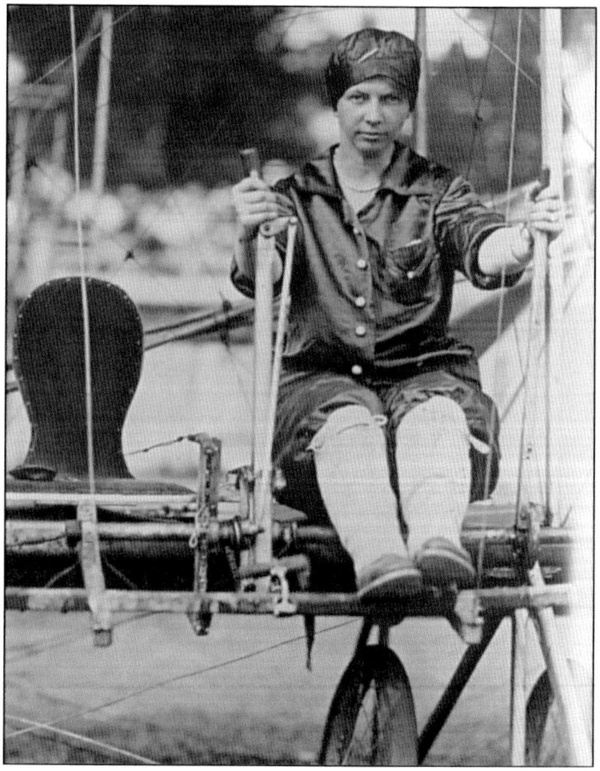

Ruth Law of Lynn gained her pilot's license in 1912 and became a stunt flyer, sometimes to the consternation of her husband. Determined to set records, she flew long-distance flights that gained her notoriety and fame, once earning her a seat at a testimonial dinner attended by none other than the president of the United States, Woodrow Wilson. (MAHS.)

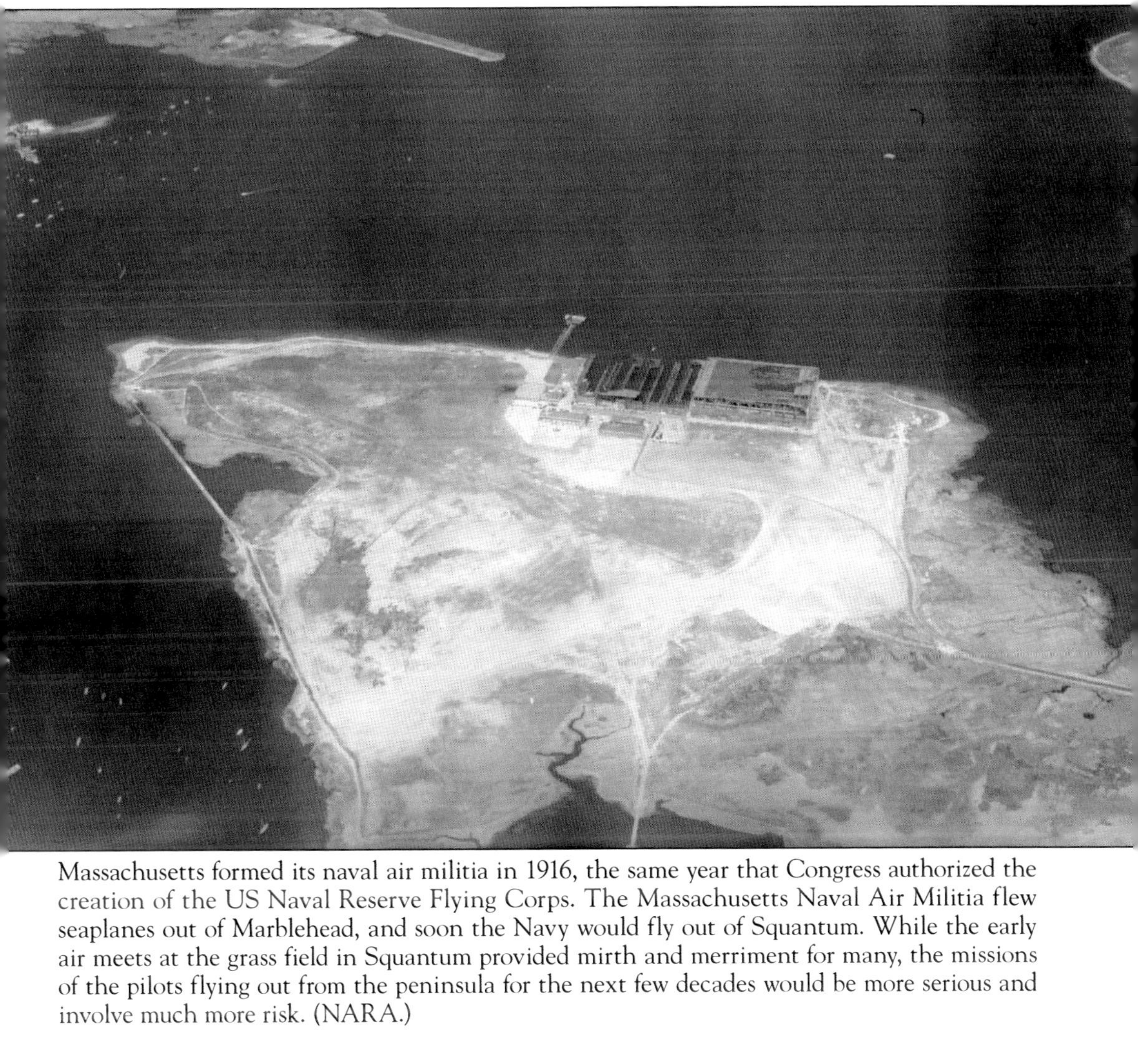

Massachusetts formed its naval air militia in 1916, the same year that Congress authorized the creation of the US Naval Reserve Flying Corps. The Massachusetts Naval Air Militia flew seaplanes out of Marblehead, and soon the Navy would fly out of Squantum. While the early air meets at the grass field in Squantum provided mirth and merriment for many, the missions of the pilots flying out from the peninsula for the next few decades would be more serious and involve much more risk. (NARA.)

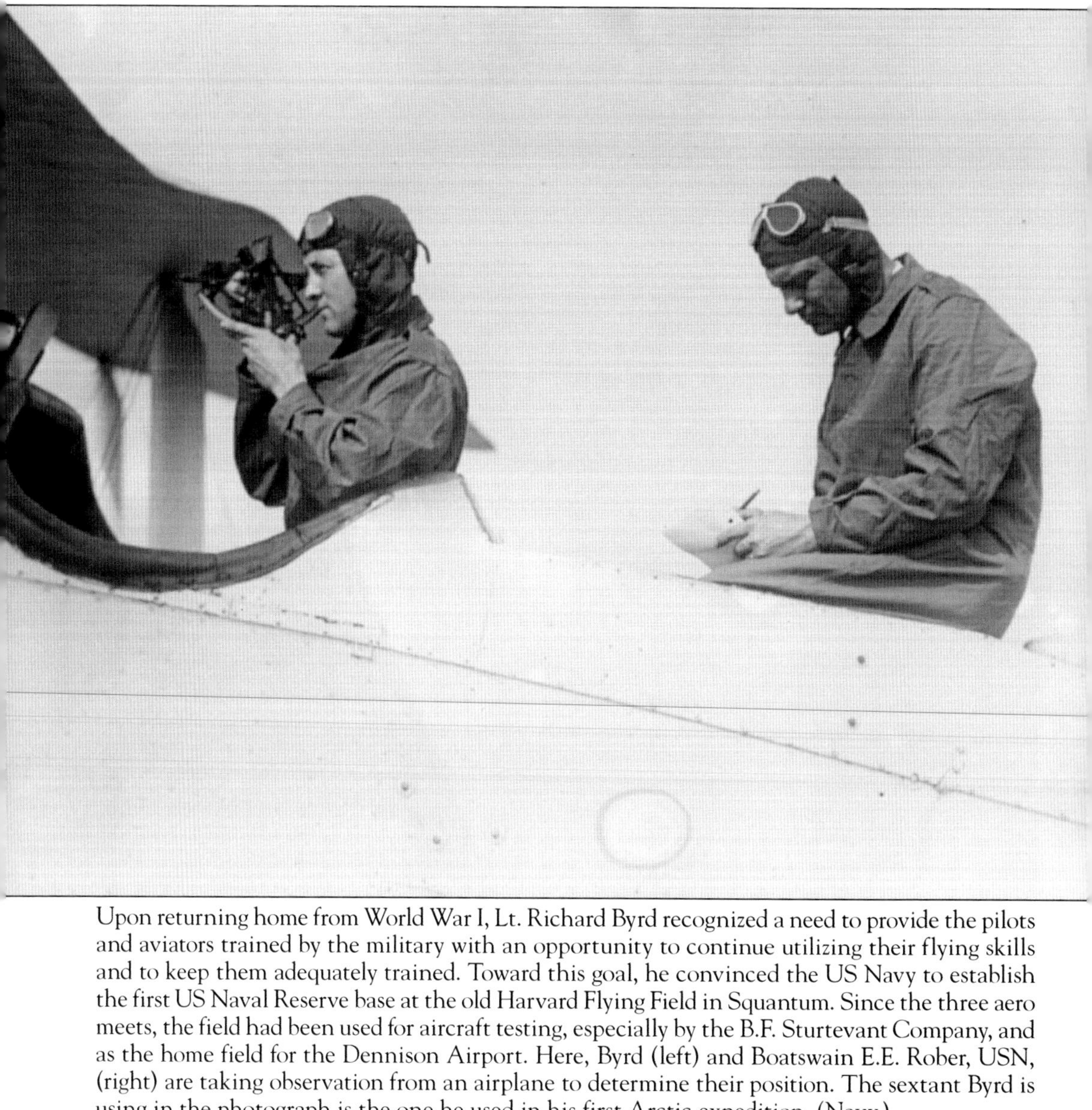

Upon returning home from World War I, Lt. Richard Byrd recognized a need to provide the pilots and aviators trained by the military with an opportunity to continue utilizing their flying skills and to keep them adequately trained. Toward this goal, he convinced the US Navy to establish the first US Naval Reserve base at the old Harvard Flying Field in Squantum. Since the three aero meets, the field had been used for aircraft testing, especially by the B.F. Sturtevant Company, and as the home field for the Dennison Airport. Here, Byrd (left) and Boatswain E.E. Rober, USN, (right) are taking observation from an airplane to determine their position. The sextant Byrd is using in the photograph is the one he used in his first Arctic expedition. (Navy.)

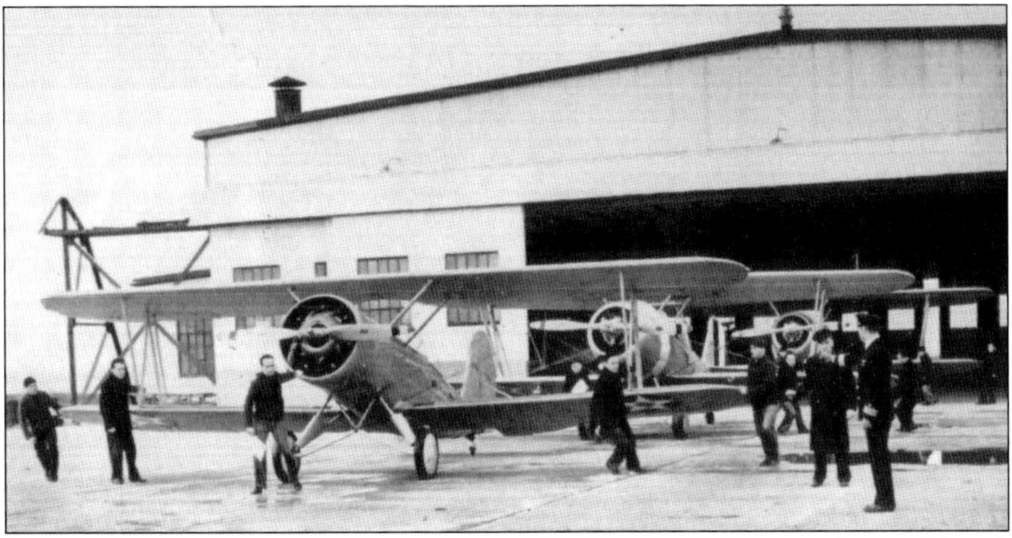

The facilities at Squantum were not designated a Naval Reserve air base until 1925, although training had been taking place there for two years by that time. Improvements to the old Victory plant facilities continued into the 1930s until sights like this one, showing crews preparing a couple of N2S Stearman planes for takeoff, became commonplace. The Stearman Kaydets, as they were officially known, worked through World War II and later became famous as crop dusters and barnstorming planes. (QHS.)

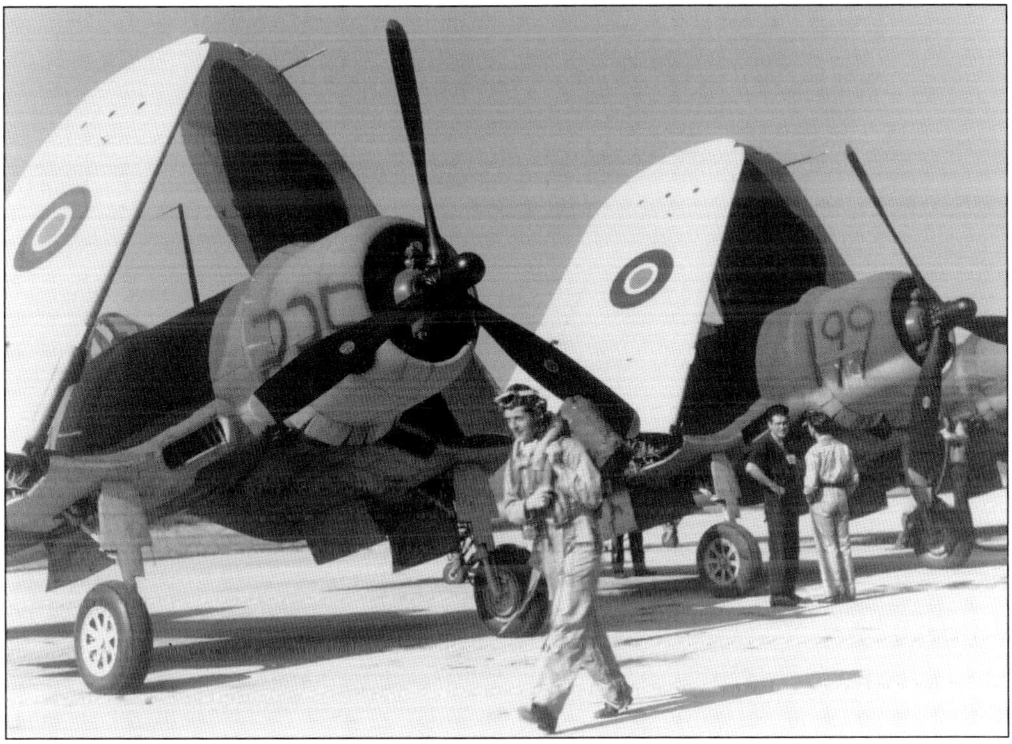

Naval Air Station Squantum would be home not only to American pilots but also, for a brief period, British. The British Fleet's Air Arm trained there from May 1943 to July 1944. Among the planes on which they trained were the Royal Navy's Vought Corsair II fighter, the F4U-1A. (NARA.)

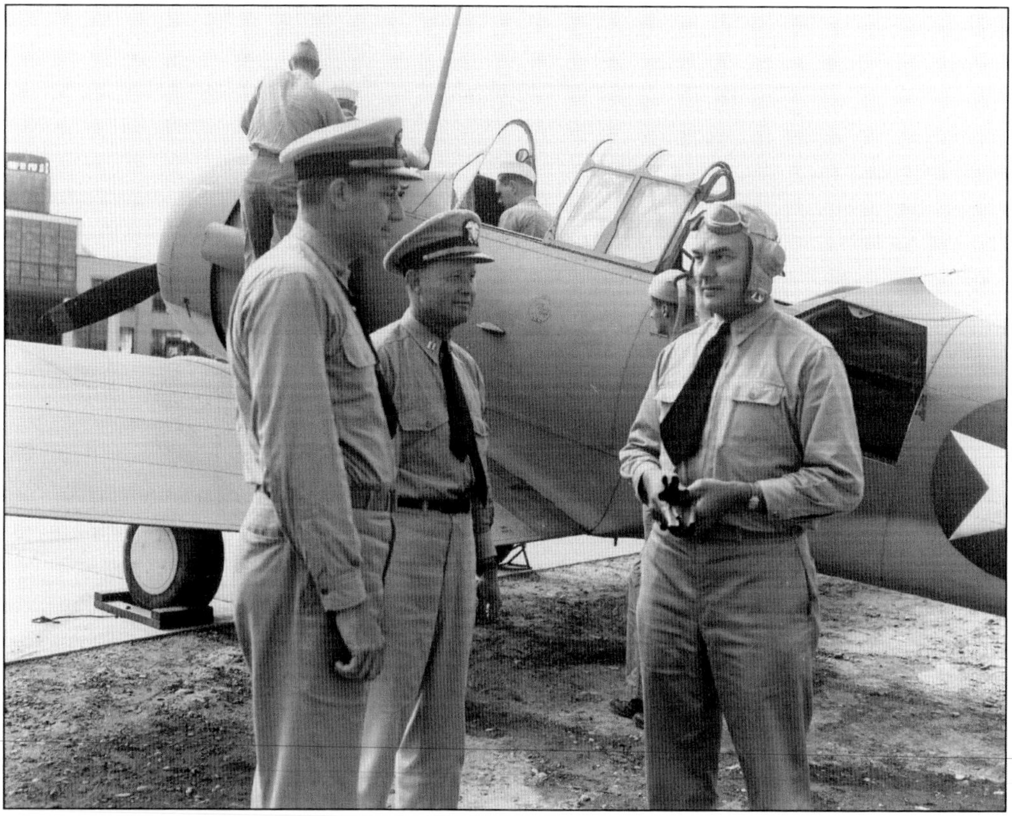

Pictured from left to right on July 23, 1942, Lt. Comdr. P.W. Watson, USN, commanding officer of Naval Air Station Squantum, and Lt. M.B. Quinn, USN, physical training instructor for the base, meet Lt. Comdr. Tom Hamilton, USN, physical training coordinator for all aviation cadets. (NARA.)

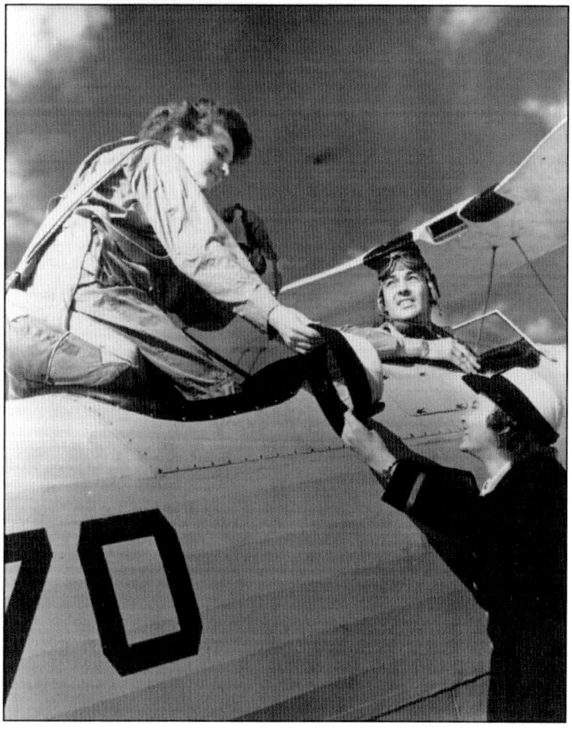

During World War II, the word "manpower" got an overhaul as it related to the war effort. At left, in October 1943, Ens. Mary McLean, USNR, a WAVE officer, hands her hat to another WAVE officer as she boards a training plane for an orientation flight at Naval Air Station Squantum. The plane appears to be an N3N Yellow Peril. (NARA.)

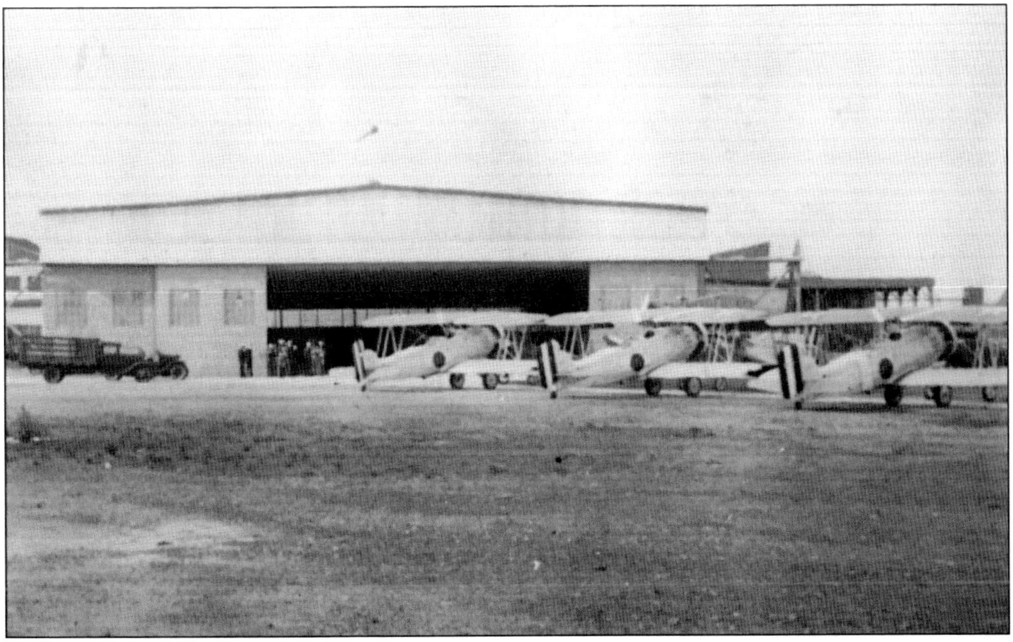

From its earliest days, the base at Squantum was home to the specific mission of training for observation. The Navy, with its focus on combat on the sea, poured its energy into antisubmarine warfare, or ASW. Among the planes on which pilots trained were Vought O2U Corsairs. (Marc Frattasio collection via Angelo Romano.)

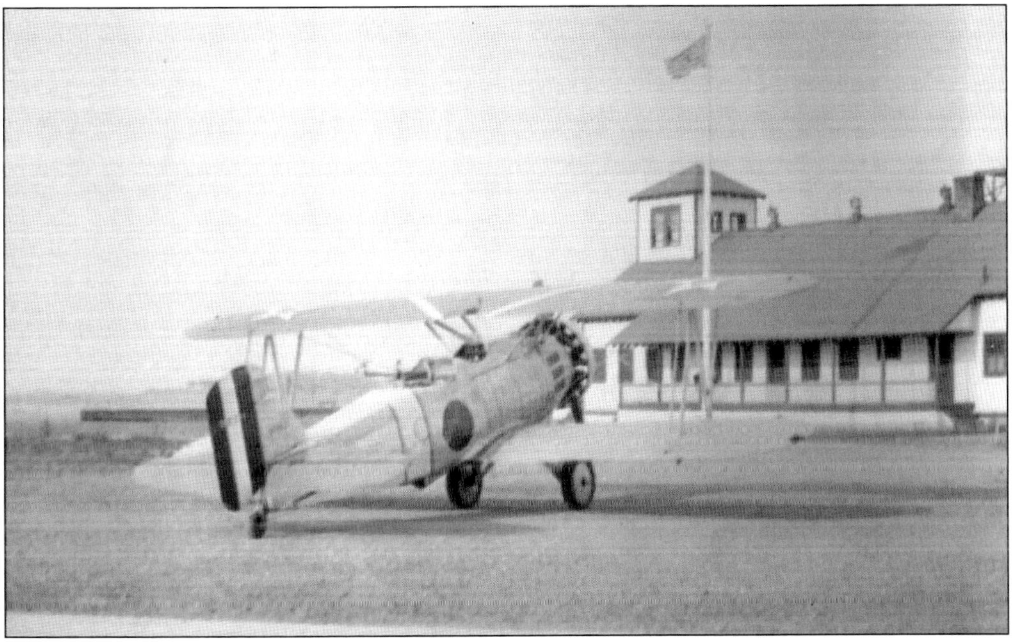

The original Curtiss F8C-1 and F8C-3 Falcons started out as shipboard Navy fighter planes in 1927 and 1928. The Marines ordered the planes as well, designating them as O2C-1 and O2C-2 and using them for observation and bombing. The Navy used the F8C-4 variant under the name that would become most famous, the Helldiver, eventually settling on the F8C-5 or O2C-1. The planes were operational through most of the 1930s. (Marc Frattasio collection via Angelo Romano.)

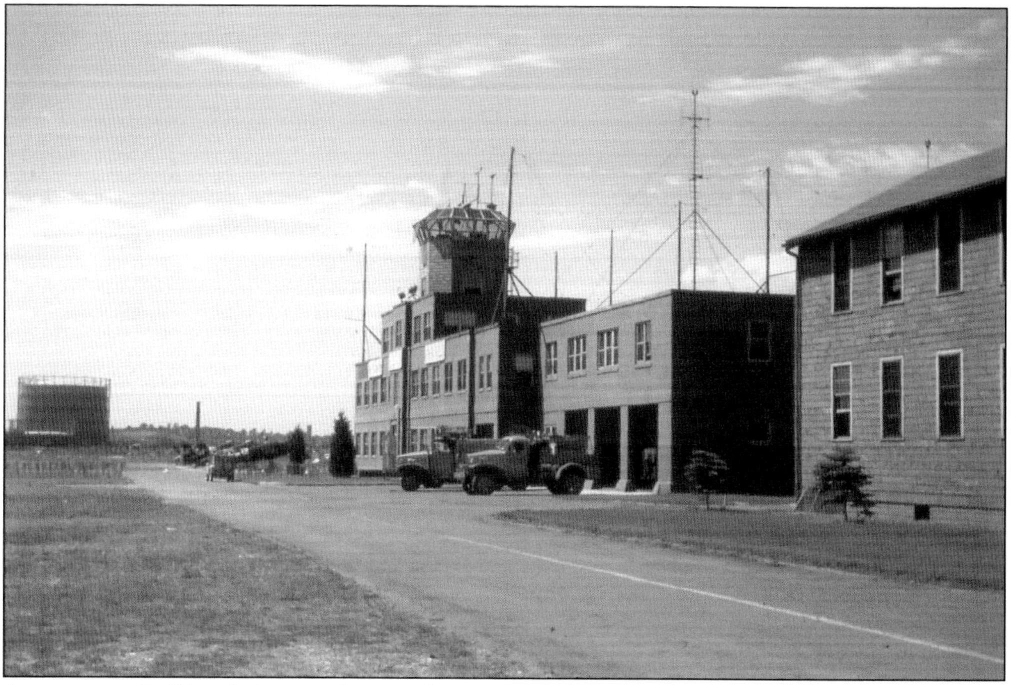

As airports of all kinds became busier, they expanded in the services they offered. Squantum's Navy base had its own fire station, a team of firefighters dedicated to watching over the planes and pilots. Control towers hit the American aviation landscape in 1930, with the first one built in Cleveland. Here, the original tower stands at center, with the old gas tank in the distance. (Marc Frattasio collection via Angelo Romano.)

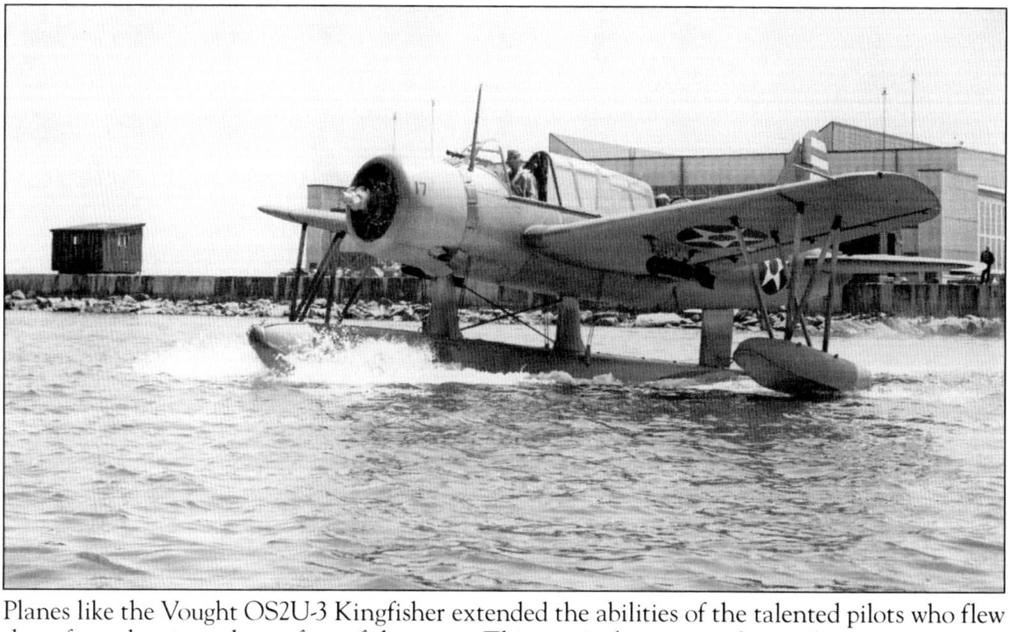

Planes like the Vought OS2U-3 Kingfisher extended the abilities of the talented pilots who flew them from the air to the surface of the water. This particular variant featured armor protection and two .30-caliber guns and had the ability to also carry either 325 pounds of depth charges or 100 pounds of bombs. Here, an OS2U-3 water taxis at Squantum. (NARA.)

Two

The Navy

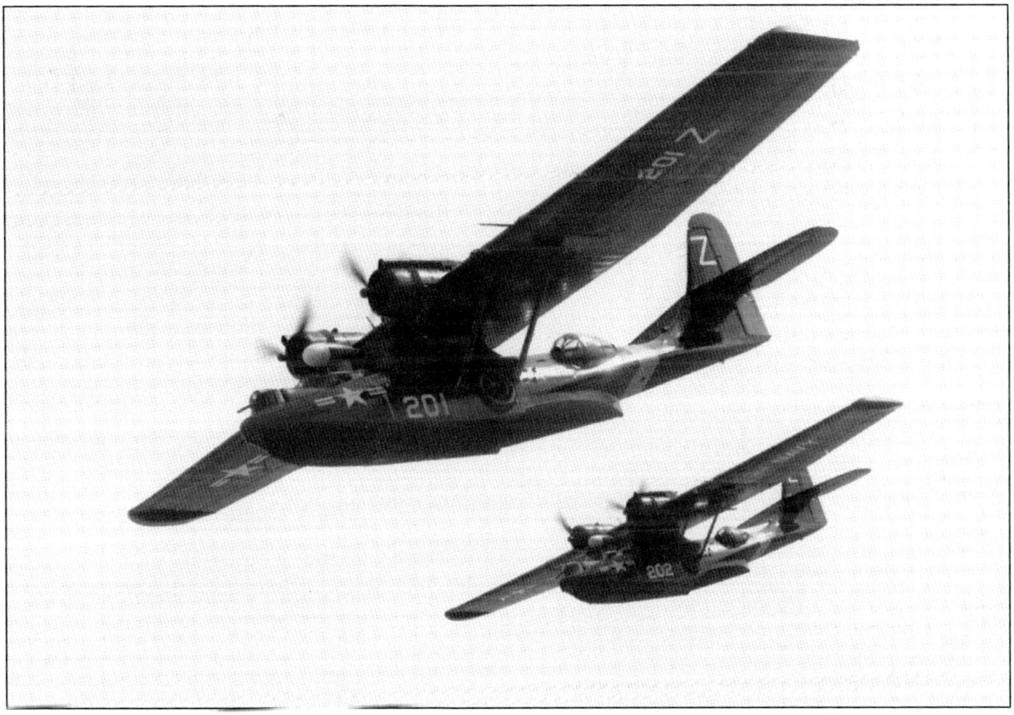

The PBY Catalina was the most extensively built amphibious aircraft in history. Aircraft manufacturers Consolidated and Douglas responded to the Navy's 1933 request to produce a long-range patrol plane with heavy cargo carrying capability, and five years later, the Navy adopted Consolidated's design as its flying boat of the future. In 1939, the service began looking for its replacement, but massive orders from allied countries at the start of World War II prompted the Navy to give the aircraft a second look. Instead of scrapping it in favor of a newer plane, the Navy placed its largest order for any plane since World War I. Used by armed forces around the globe, the "Cat" played many roles in many battles. This particular plane, a PBY-5A, was given a retractable tricycle undercarriage for use in amphibious operations. Slow and unarmored, the Catalina was vulnerable when attacked, but it carried out nighttime raids for the Navy and performed superbly in search and rescue operations. (Navy.)

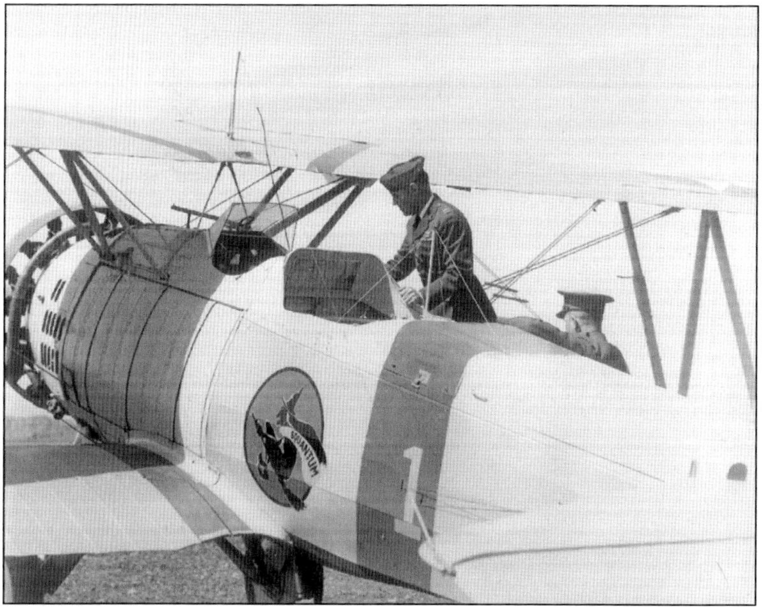

The men stationed at Squantum selected the "Squantum Witch" for their insignia, citing the history of the Massachusetts Bay Colony. The image of the witch riding her broom into the sky adorned the base newspaper and even the bodies of the planes themselves, as shown here on an N3N Yellow Peril trainer. (Navy.)

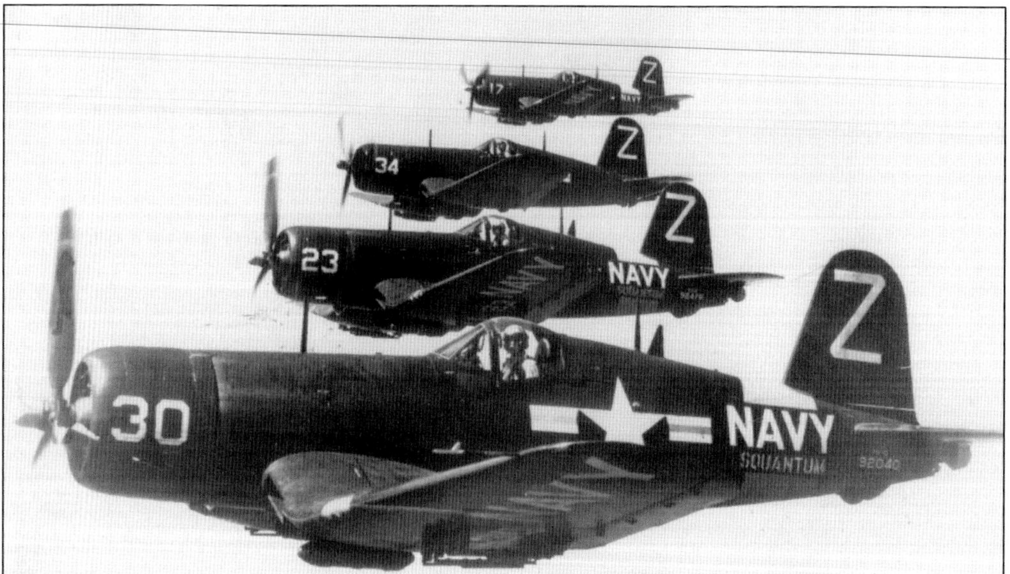

If one could not get close enough to this formation of Vought F4U Corsairs to read the word "Squantum" under the tail section, all one had to do was read the large letter "Z" on the tail to know where these planes had come from. This classic shot of classic planes was used in air show posters a half century after the photograph was taken. (Navy.)

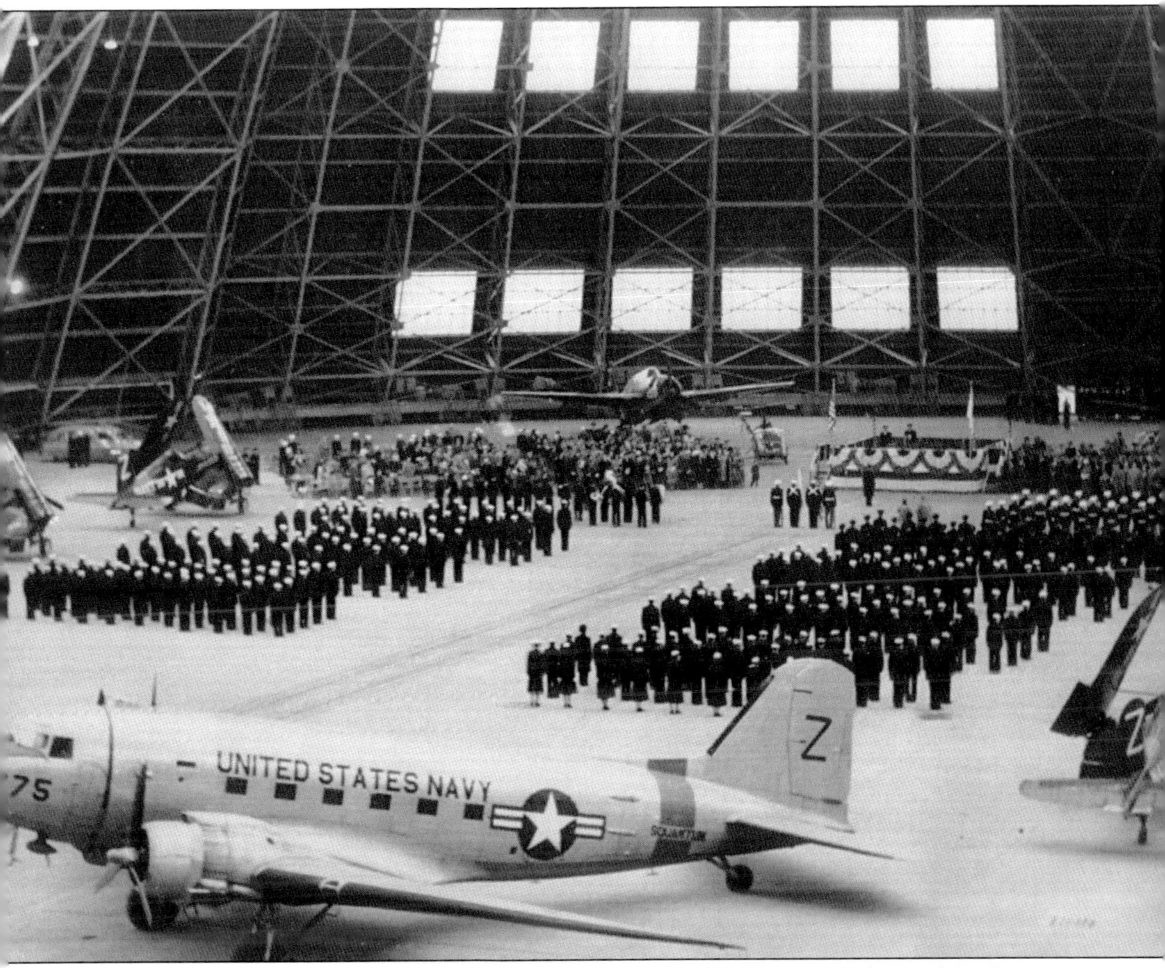

With the growth, expansion, and heavy utilization of the newer South Weymouth Naval Air Station through the 1940s and 1950s and the growth of Boston's Logan Airport causing an increase in air traffic over Boston Harbor, the Squantum base was closed and its units transferred to South Weymouth. Squantum was designated as an active naval air station (no longer a reserve base) on March 5, 1942, and closed on December 4, 1953. Operations transferred from the water's edge to the inland base, home to two hangars including the massive Hangar 1. The men who worked at the base claimed that the mammoth structure produced its own weather. Condensation dripping from the ceiling fell like rain, and fog often formed like clouds in the upper reaches. Here, an R4D Skytrain stands in the foreground as the Squantum men are welcomed to South Weymouth. Even the Squantum "Z" made the transition, now becoming the designating mark of the South Weymouth station. (Navy.)

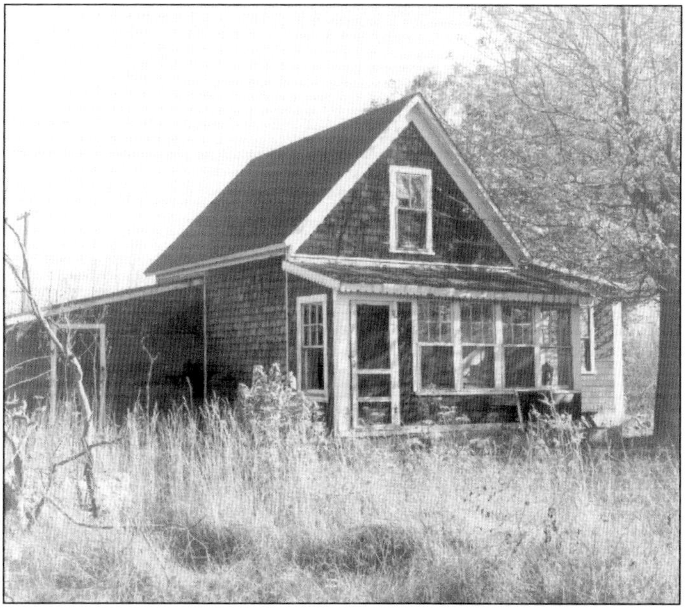

To build the naval air station at South Weymouth (and, technically, parts of Rockland and Abington), the Navy purchased numerous house lots in a section of Weymouth known as the Old City, mostly a wooded area of pig farms. The settlement went back to Colonial days, but in the late 1930s, it was still a place of unimproved roads and, according to the July 2, 1942, *Weymouth Gazette and Transcript*, "disastrous fires, an unsolved murder and gathering of uncounted bushels of berries." (NARA.)

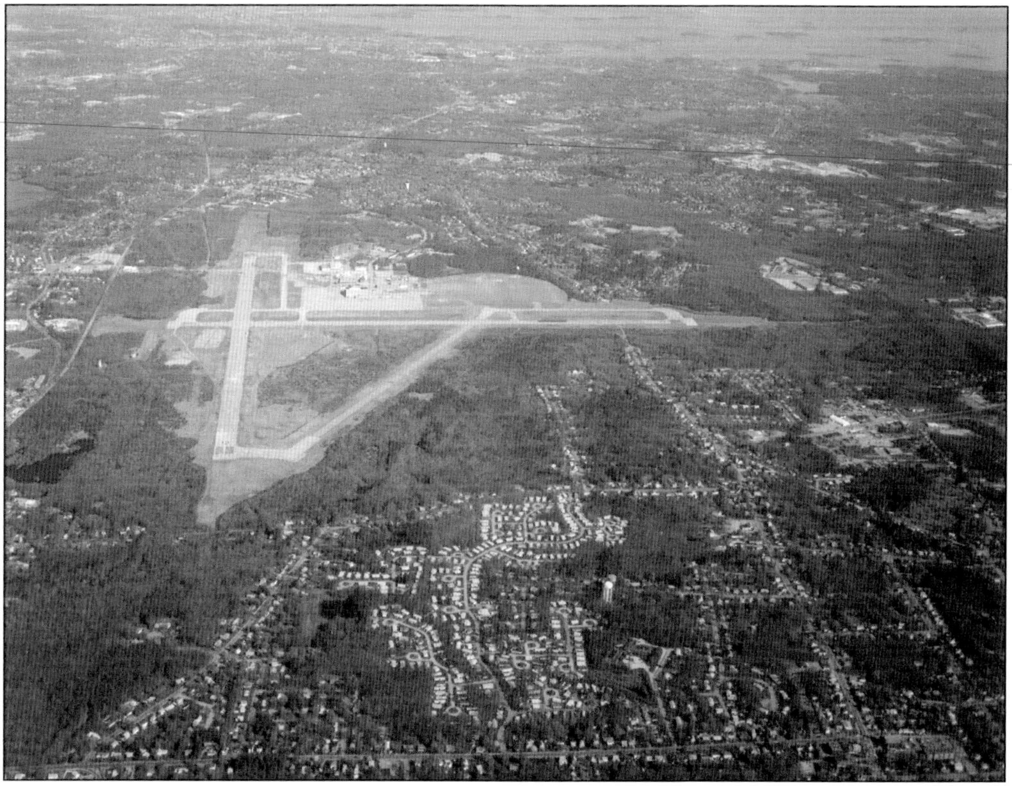

Local pilot and aviation proponent Ronald White arranged to take Navy commander Charles Rosendahl and Massachusetts aeronautics commissioner Crocker Snow on an aerial tour to survey the area for the Navy lighter-than-air facility proposed for the local area. The fruit of that labor would eventually be runways for helicopters, jets, and more, adaptable from the ASW missions of World War II to Cold War readiness in the 1950s and beyond. (Navy.)

Hangar 1 was a marvel, constructed to hold six blimps. It was made of steel and covered eight acres. The dimensions were 956 feet long, 294 feet wide from outside wall to outside wall, and 191 feet high. The doors were known as "orange-peels" and weighed 250 tons each. It took four large electric motors to open and close each door along railroad tracks. (NARA.)

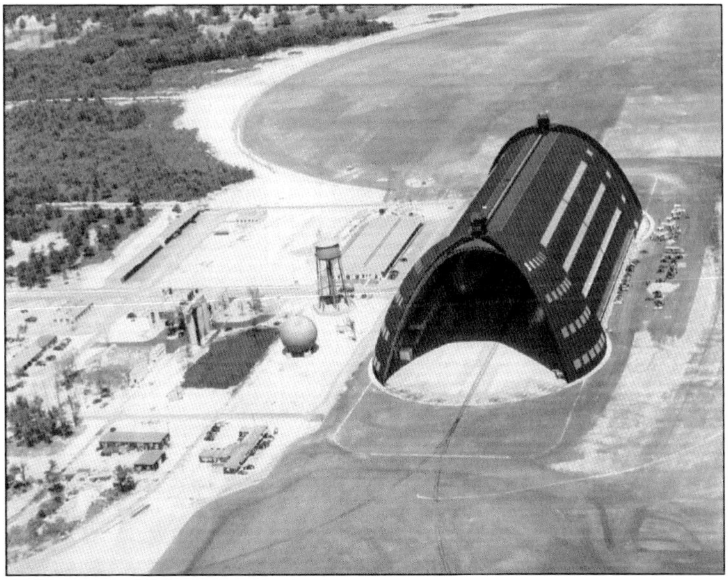

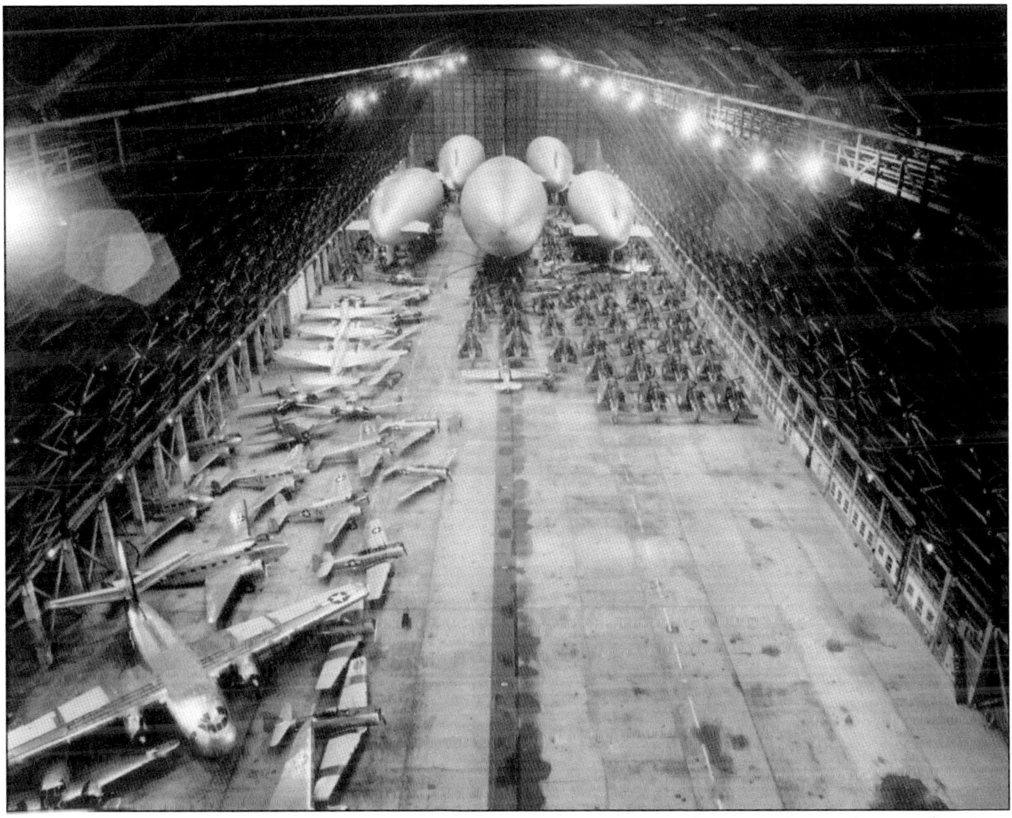

When severe storms threatened the coast, planes and blimps alike sought shelter and safety. This photograph shows about 90 fixed-wing aircraft and blimps all sheltered in Hangar 2 during a storm. Most likely, this image was taken just before or during the Great Atlantic Hurricane of September 14, 1944, which claimed the Vineyard Sound Lightship and its crew among its casualties. (NARA.)

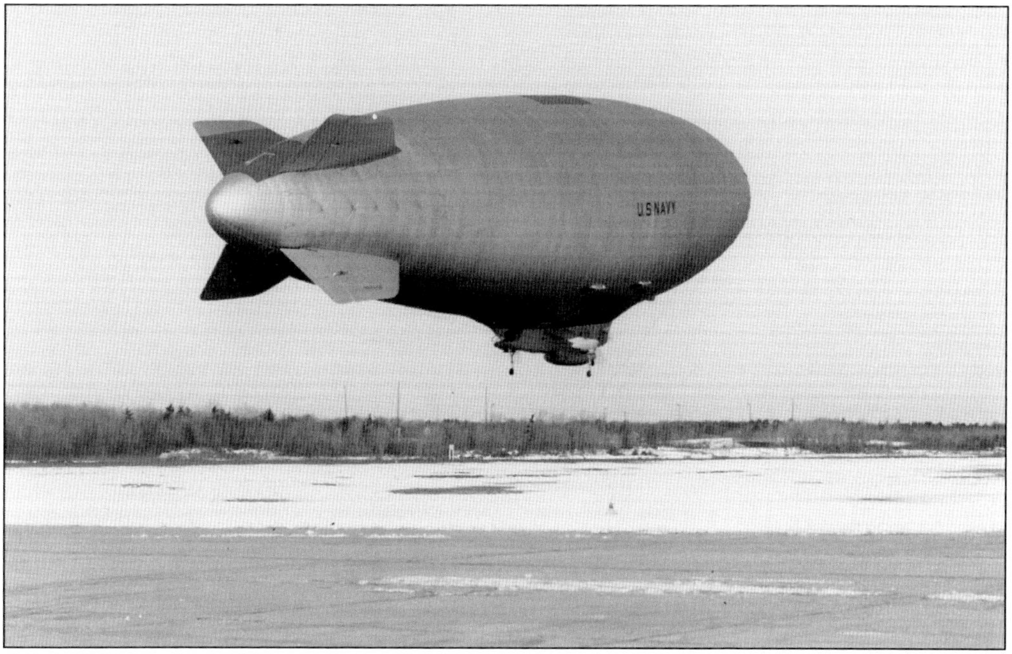

South Weymouth would become famous for its LTA craft, or, more popularly stated, its blimps. Here, the ZPG-2 "Snowbird" departs the station on March 4, 1957, to float into history. Comdr. Jack R. Hunt crisscrossed the Atlantic twice, touching down at Naval Air Station Key West after 11 days and covering 9,448 miles without stopping for fuel. (NARA.)

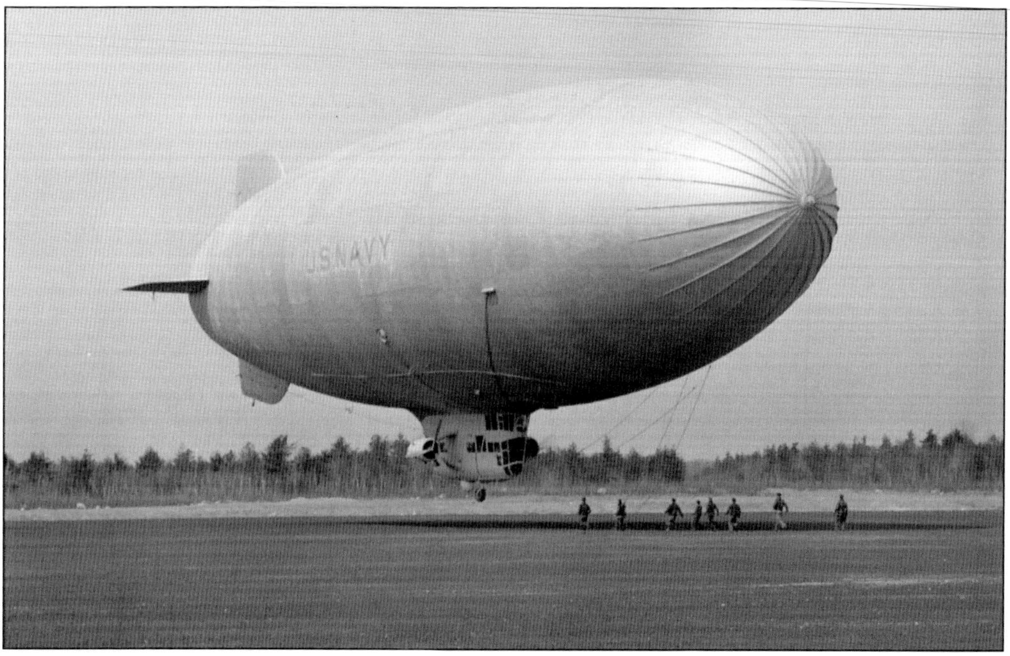

Here, on March 17, 1945, the K-82 lands at South Weymouth with an inspection party. K-class airships grew out of the Depression. The Navy ordered 134 of them, eventually using them in ASW roles during World War II. Debates raged at the time over whether or not blimps or seaplanes would be the future for the mission. (NARA.)

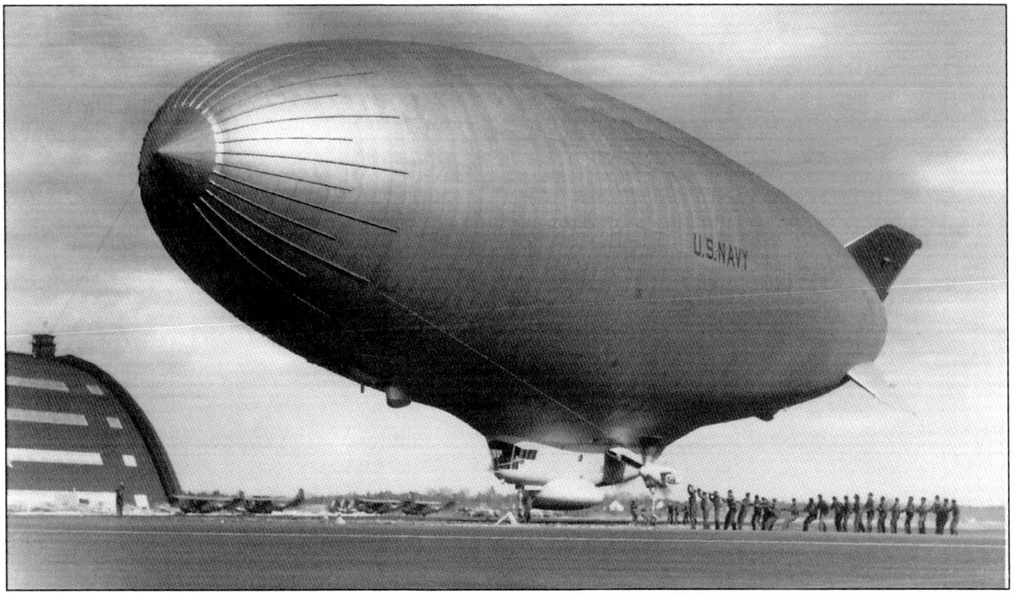

A total of 18 N-class blimps were developed for the Navy by the Goodyear Aircraft Company of Akron, Ohio, including 12 ZPG-2s. The blimps measured 343 feet in length and 107 feet in height with a diameter of 76 feet. They could travel at a maximum speed of 80 miles per hour and had endurance of more than 200 hours. (Navy.)

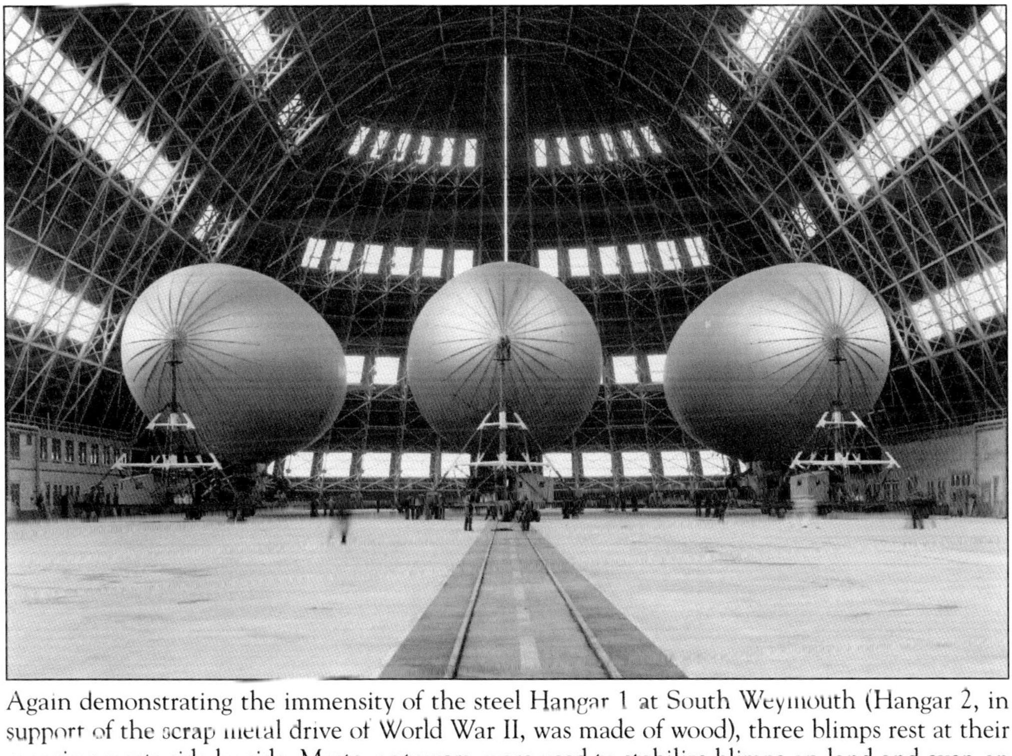

Again demonstrating the immensity of the steel Hangar 1 at South Weymouth (Hangar 2, in support of the scrap metal drive of World War II, was made of wood), three blimps rest at their mooring masts side by side. Masts, or towers, were used to stabilize blimps on land and even on ships. The technology quickly followed the development of lighter-than-air craft, first used in England in 1911. (NARA.)

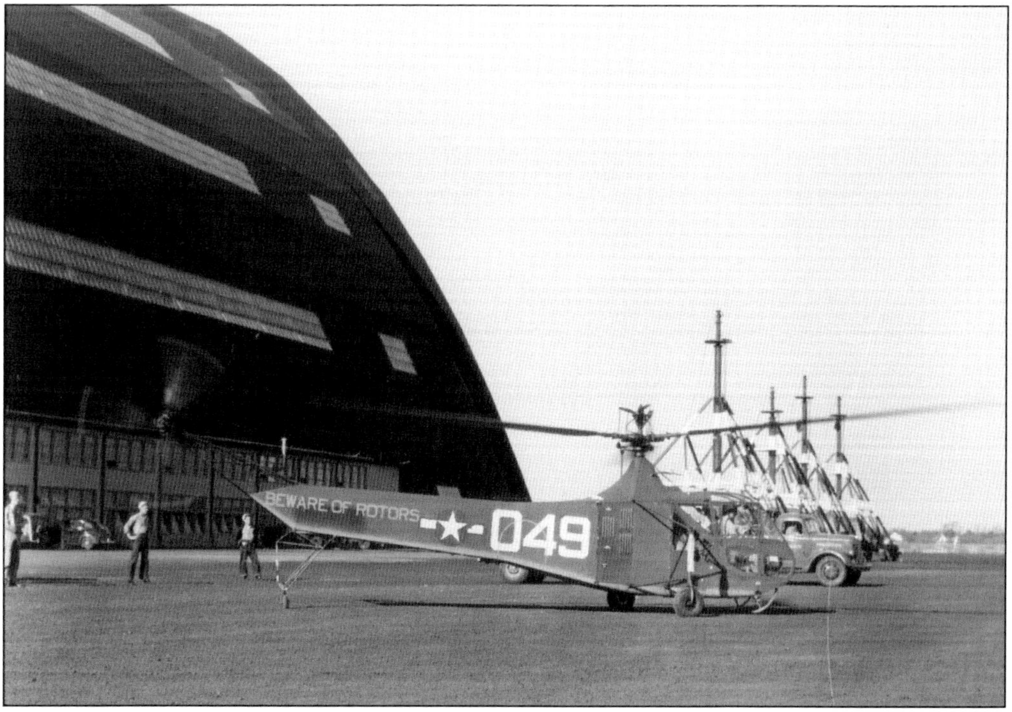

The first helicopter at South Weymouth, around 1952, was a Hiller HTE-2 used for training and utility purposes. Helicopters saw an increase in use and experimentation during World War II, and they became widespread in the 1950s and 1960s, featuring in the Korean and Vietnam wars. ASW helicopters were attached to the base until the late 1960s. The SH-3A Sea King helicopter was one of the last to be used in this work at this base. (NARA.)

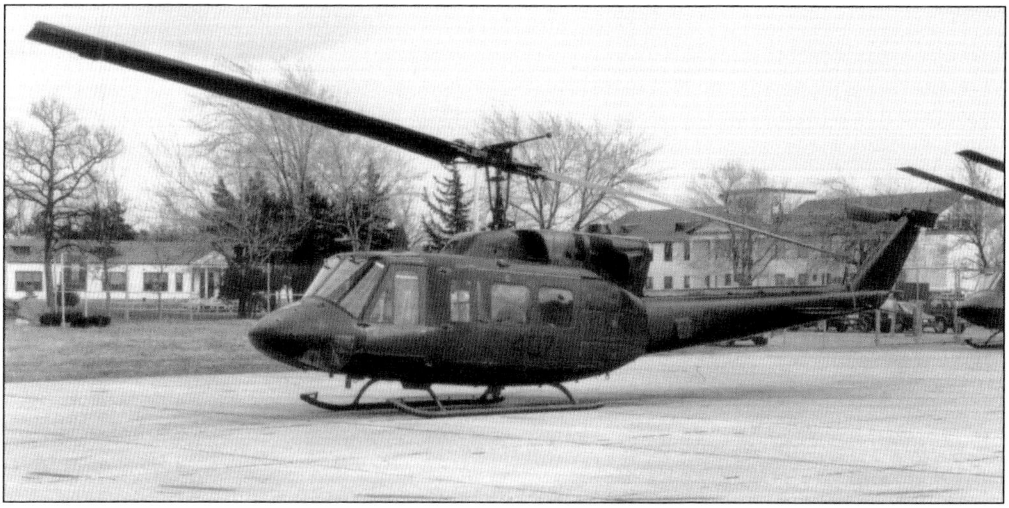

Approximately 40 countries around the world use the Bell UH-1 Iroquois helicopter, better known as the "Huey." Since the 1950s, more than 9,000 Hueys have been manufactured, the first American versions seeing combat in Vietnam in 1963. Marine Light Helicopter Squadron 771, based at South Weymouth from the 1950s until its deactivation on August 1, 1994, flew them. Flexible to fit all sorts of mission requirements, it may be the most widely used helicopter in the world. (Navy.)

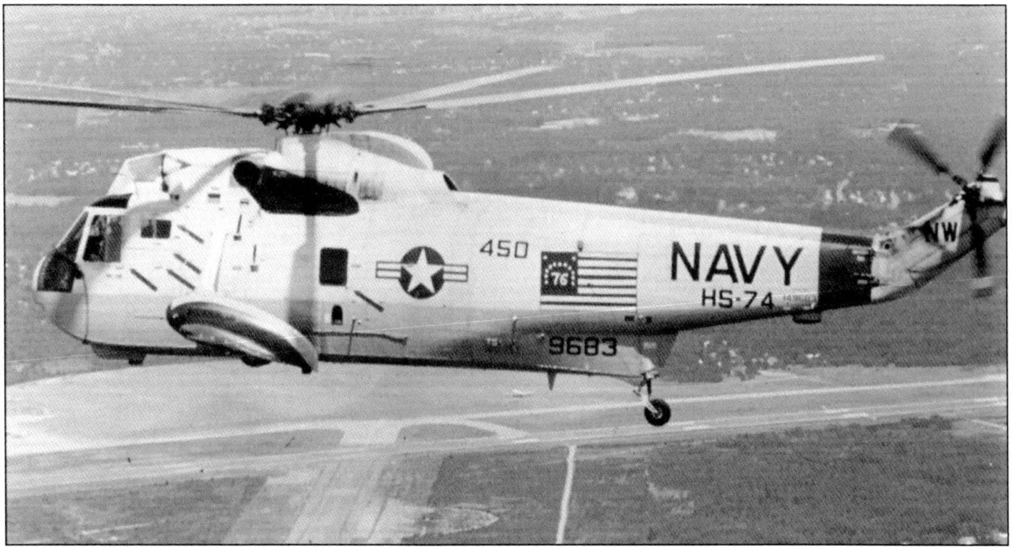

Helicopter Antisubmarine Squadron 74, or HS-74, flew out of South Weymouth from 1973 to 1984 using the SH-3 Sky King. It later transitioned into HSL-74 and flew SH-2F Seasprite variants from 1985 to 1994. Primarily used for antisubmarine warfare, Sikorsky's Sea King was the world's first amphibious helicopter. Introduced in 1961, it was retired from Navy service in 2006. (Navy.)

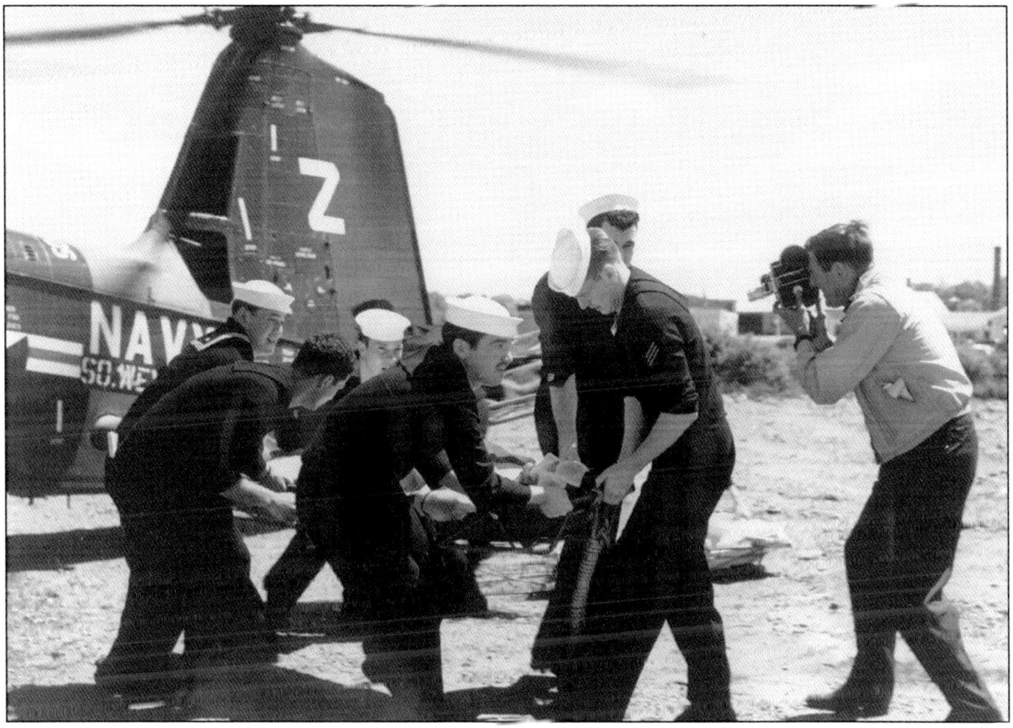

On May 26, 1954, at 6:11 a.m., fluid from a hydraulic catapult aboard the aircraft carrier USS *Bennington* leaked and was ignited by the flames from a jet, causing a massive explosion that killed 103 crewmen and injured 201 others. Just off Rhode Island's Narraganset Bay, the ship proceeded under her own power for port. Here, a helicopter from Naval Air Station South Weymouth delivers an injured crewman to the naval hospital in Newport, Rhode Island. (Navy.)

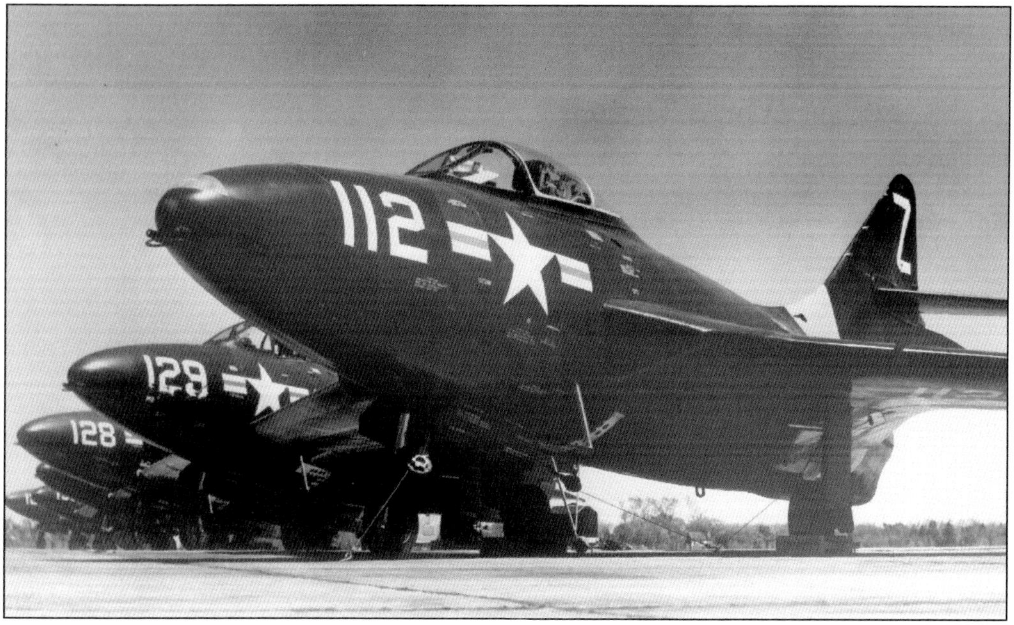

Through the years, Naval Air Station South Weymouth witnessed numerous styles of aircraft, moving from blimps and prop planes to helicopters and jets. Here, a lineup of F9F-6 Cougars (which had swept wings, as compared to the F9F-5, which had straight wings) stands at the ready at the base. Residents of the surrounding towns became used to the sight of low-flying Navy aircraft. Today, it would be an anomaly. (Navy.)

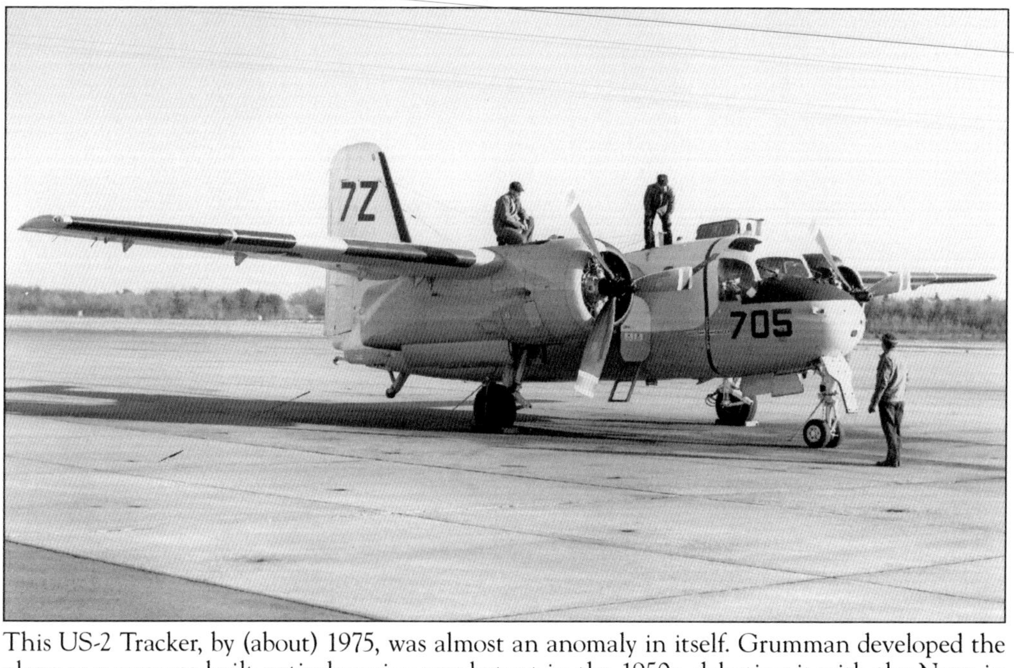

This US-2 Tracker, by (about) 1975, was almost an anomaly in itself. Grumman developed the plane as a purpose-built antisubmarine combatant in the 1950s, debuting it with the Navy in February 1954. The Tracker combined the capabilities of its predecessor, the Grumman AF Guardian, which came in two versions, one with detection equipment, the other with weaponry. The Tracker carried both. (Navy.)

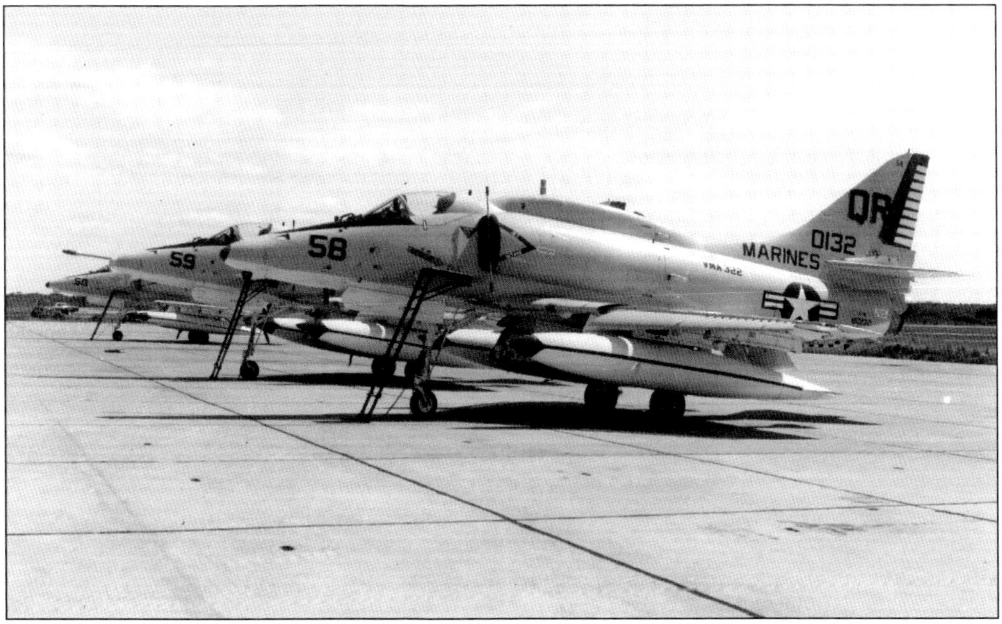

Marine Attack Squadron 322 (or VMA-322) was formed in 1943 and remained active through the end of November 1949 in the Pacific. Revived in 1951, the unit moved to South Weymouth three years later and remained there until deactivated for good on June 27, 1992. The unit flew F4U Corsairs, F9F Cougars, FJ3 Furies, and finally, shown here, A-4 Skyhawks, from 1962 to 1992. A preserved Skyhawk stands as the "gate guardian" of the old base today. (Navy.)

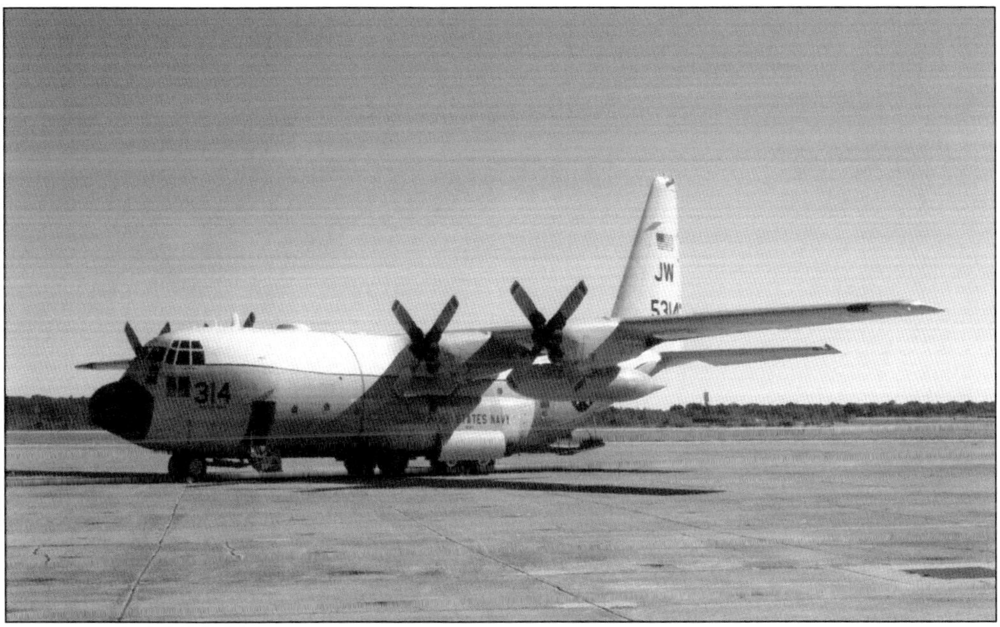

Fleet Logistics Support Squadron 62, or VR-62, flew C-130T Hercules transport planes out of South Weymouth. The transport can—and still does—carry as many as 92 ground troops or 64 paratroopers and still have room for equipment. If necessary, it can carry 74 patients on stretchers with attendants. It has also been used for search and rescue, scientific research support, weather reconnaissance, aerial refueling, and aerial firefighting. (Marc Frattasio collection via Angelo Romano.)

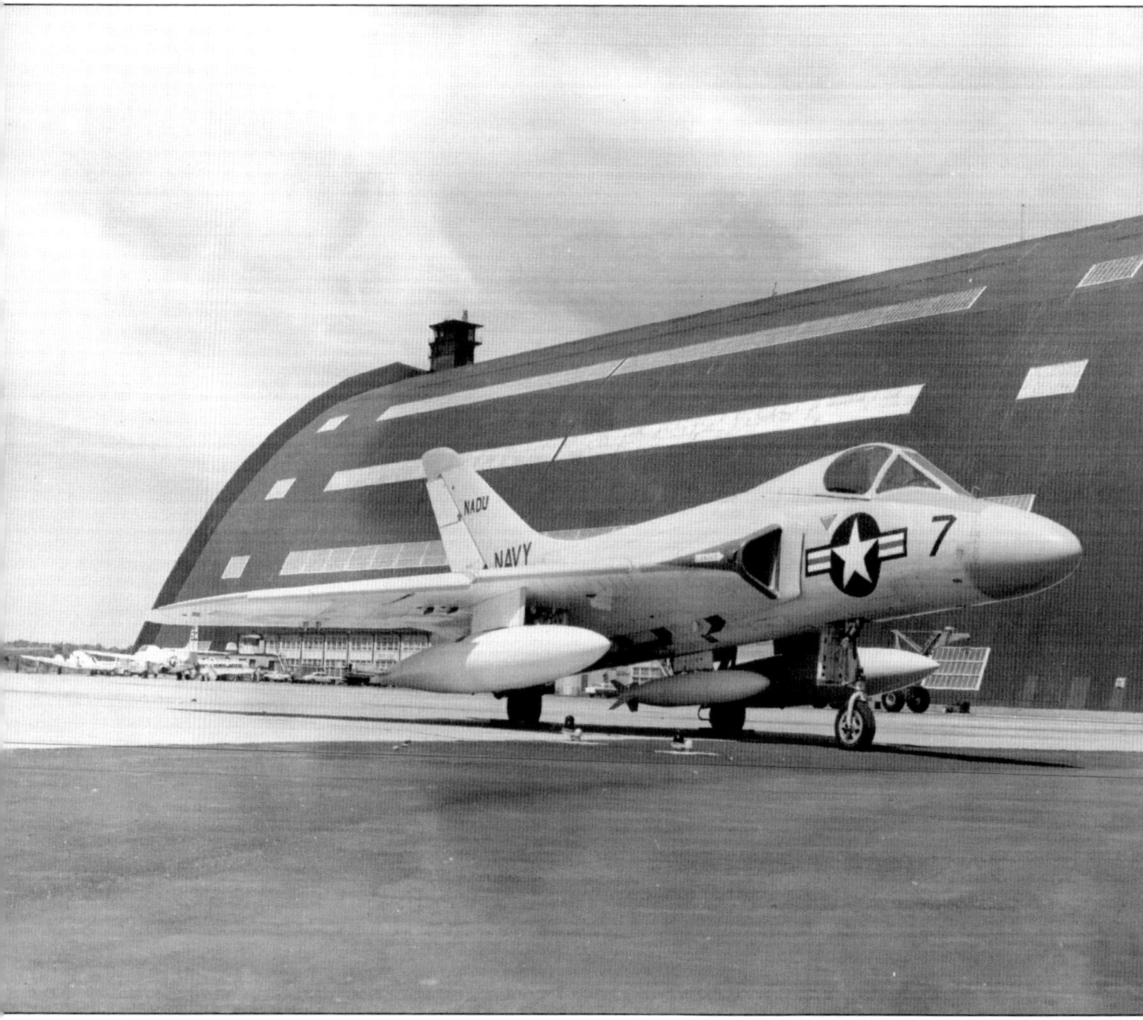

One of the most interesting yet least studied units flying out of South Weymouth, the Naval Air Development Unit, or NADU, operated on base from 1953 to 1961. This secretive outfit provided flight testing support for projects initiated at the MIT Lincoln Laboratories and by other defense contractors. NADU utilized numerous different types of aircraft as it tested experimental electronic equipment and other systems having to do with the bane of the day, the enemy submarine. Aircraft included the ZPG-2W blimps (seen on pages 50 and 51), Douglas F3D Skyknights, Douglas F4D Skyrays, Lockheed P2V Neptunes and Lockheed WV-2 Warning Stars. Here stands an F4D Skyray at the ready. According to historian Marc Frattasio, "These were used as interceptors to test the SAGE air defense system . . . the WV2s and ZPG2 blimps were used to pick up radar contacts and develop tracks that were fed into the SAGE computer. The computer then plotted an intercept solution and transmitted it to the interceptor, which then intercepted the radar contact." (Navy.)

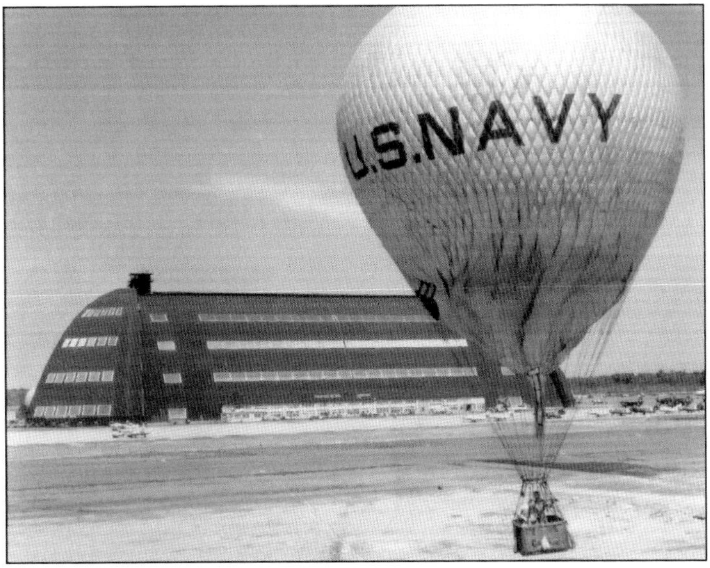

In July 1957, NADU acquired a free balloon of 40-foot diameter to test electronic equipment from the Lincoln Labs. The balloon could be set up to be free of magnetic fields, an important consideration in the research being conducted. A crew of three went aloft, with a 20-man ground crew in trucks below. In June 1958, it caused an uproar in the town of Weymouth's Columbian Square, making an unexpected landing. (Navy.)

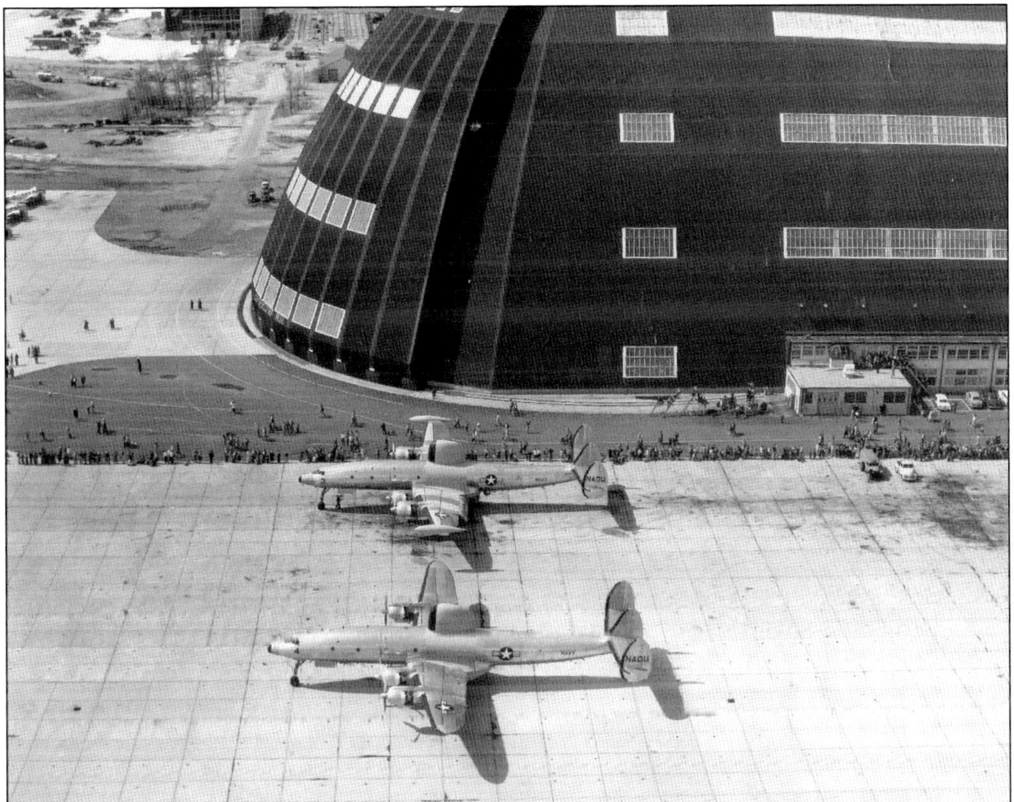

The primary role of NADU's ZPG-2 blimps and the WV-2 Warning Stars was to pick up radar contacts that would be sent back to the SAGE computer system for relaying an interception route to the F4D Skyrays. Lockheed's Warning Stars served both the Navy and the Air Force in radar surveillance and intelligence gathering. (Navy.)

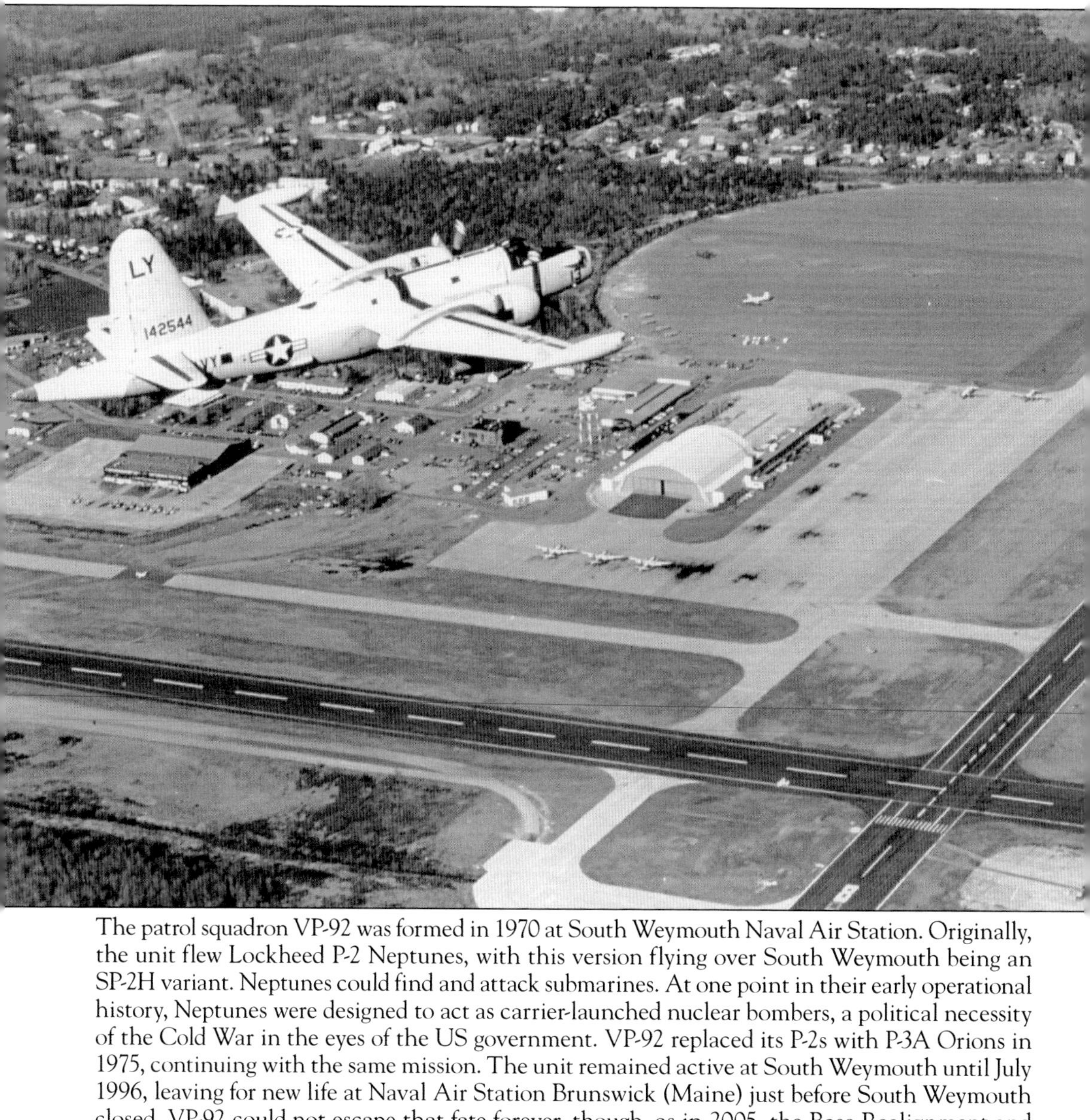

The patrol squadron VP-92 was formed in 1970 at South Weymouth Naval Air Station. Originally, the unit flew Lockheed P-2 Neptunes, with this version flying over South Weymouth being an SP-2H variant. Neptunes could find and attack submarines. At one point in their early operational history, Neptunes were designed to act as carrier-launched nuclear bombers, a political necessity of the Cold War in the eyes of the US government. VP-92 replaced its P-2s with P-3A Orions in 1975, continuing with the same mission. The unit remained active at South Weymouth until July 1996, leaving for new life at Naval Air Station Brunswick (Maine) just before South Weymouth closed. VP-92 could not escape that fate forever, though, as in 2005, the Base Realignment and Closure Commission decided to not only close Brunswick Naval Air Station but to disestablish the unit for good. The squadron was disestablished on October 17, 2007. (Navy.)

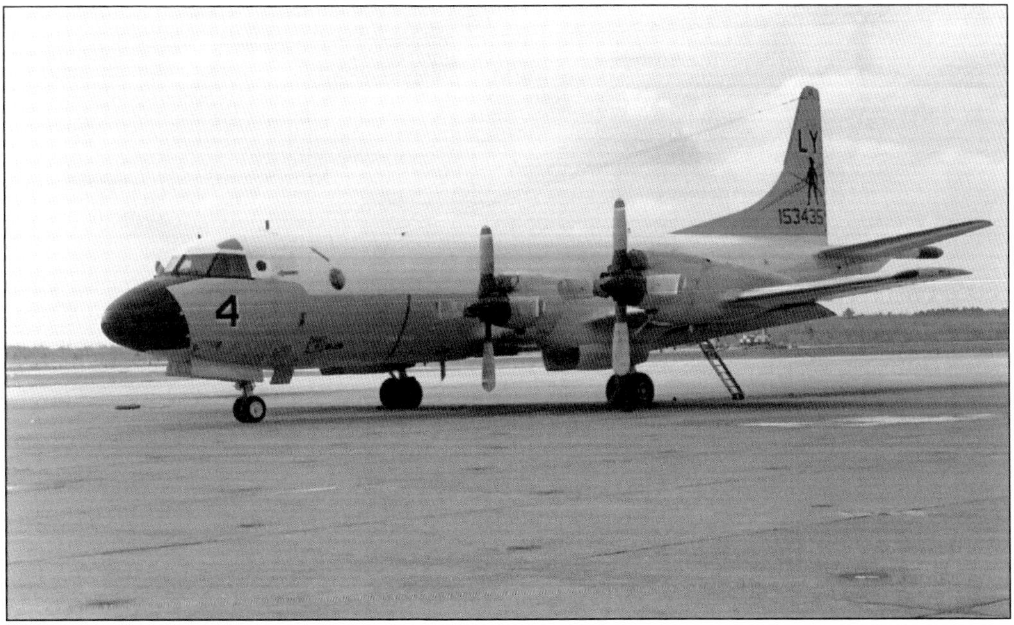

VP-92's P-3s became a familiar sight to the people of the towns surrounding Naval Air Station South Weymouth. They were, in fact, the symbolic workhorses of the base. Keeping up regular rotations of antisubmarine and maritime surveillance patrols, they became trusted allies of the people of Weymouth, Abington, and Rockland. (Navy.)

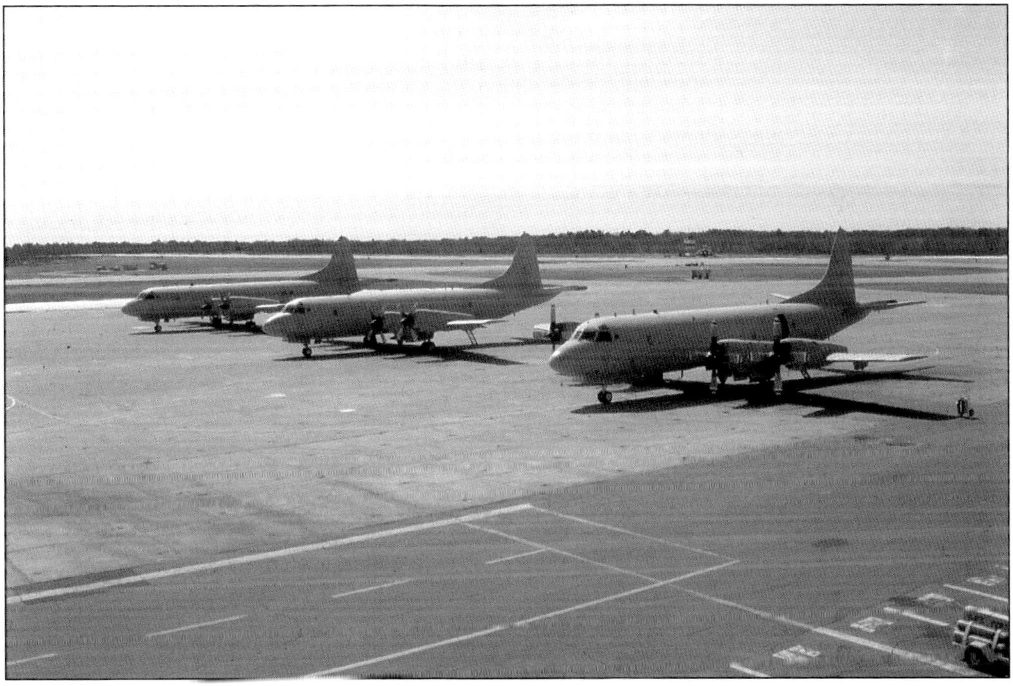

The P-3 (a P-3B above, a P-3C here) was fashioned on the airframe of Lockheed's Electra airliner (L-188). The plane can maintain a 13-hour mission and be armed with torpedoes, mines, and depth bombs, and is still, barely, in use in the United States. They are currently being replaced by P8-A Poseidons. (Marc Frattasio collection via Angelo Romano.)

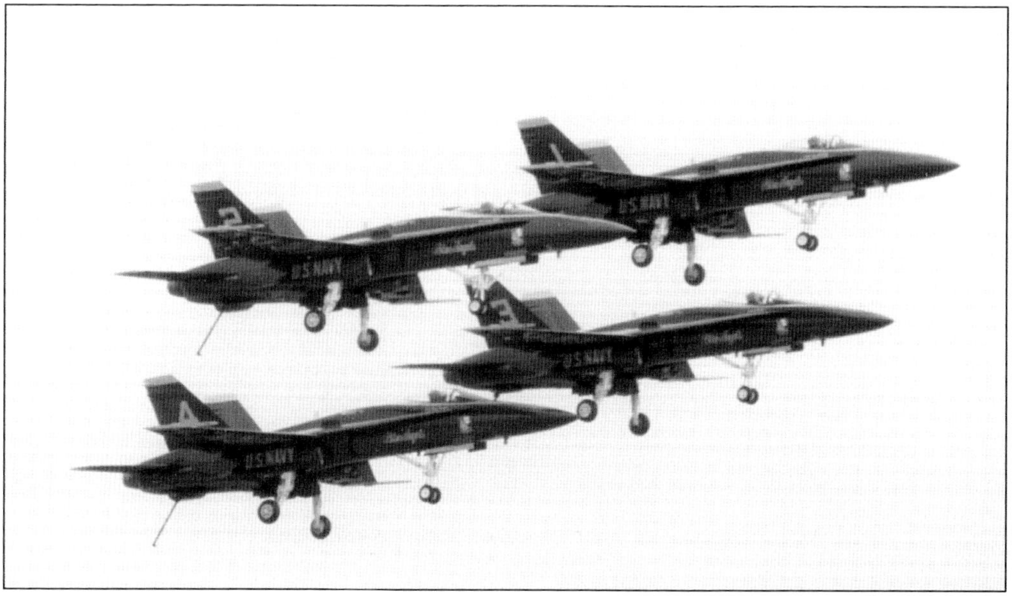

Naval Air Station South Weymouth became a gathering place for aviation enthusiasts through its high-powered air shows. The first Blue Angels flight demonstration team flew the Grumman F6F Hellcat but quickly changed within the first year to the F8F Bearcat for its power and maneuverability. In 1949, the Blue Angels transitioned into the jet age and have since flown the F9F-8 Cougar, F11F-1 Tiger, F-4J Phantom and others before settling into today's F/A-18 Hornets, which they've been using since 1986. (Navy.)

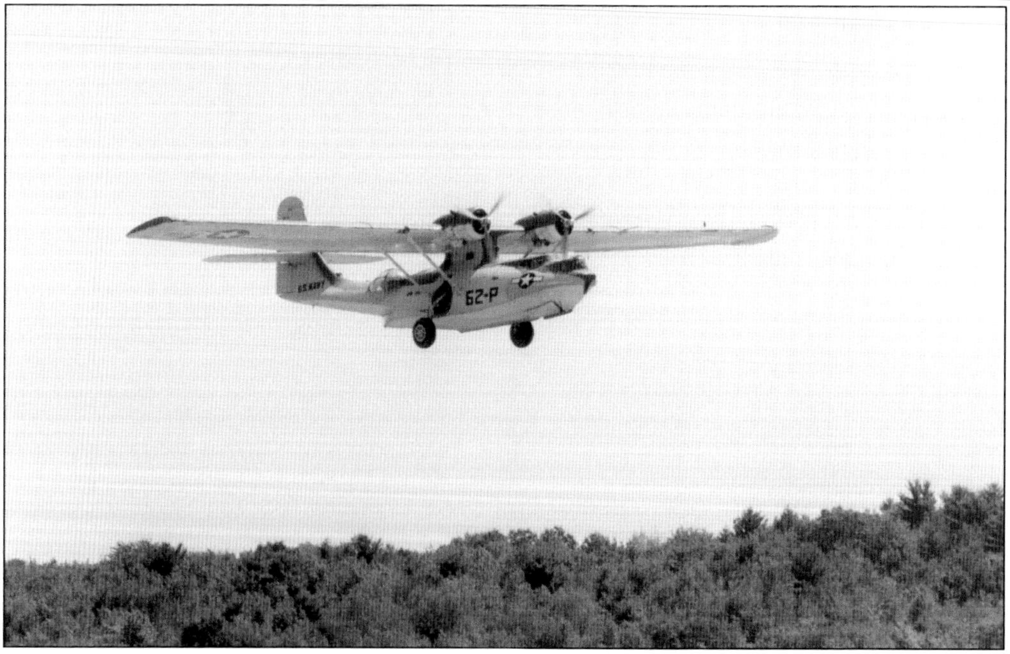

The air shows offered a chance for spectators to see living history, as the planes of yesteryear shared the skies with modern craft. The Consolidated PBY Catalina was the answer to America's need for a long-range patrol boat aircraft in the late 1930s and served with distinction throughout World War II in almost every theater of action. Old-timers remembered them from Squantum. (Navy.)

The AV-8B Harrier II, a vertical takeoff and landing airplane, was also popular during the shows. The Harrier II saw heavy action in Operation Desert Storm as the first Marine Corps tactical strike aircraft into the theater of operations and, based only 35 miles from the Kuwait–Saudi Arabia border, flew the shortest distance into combat of any aircraft. (Navy.)

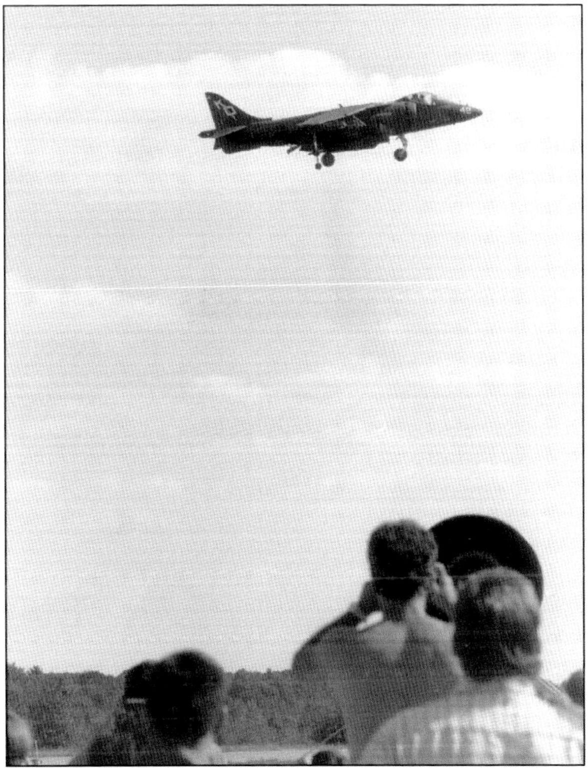

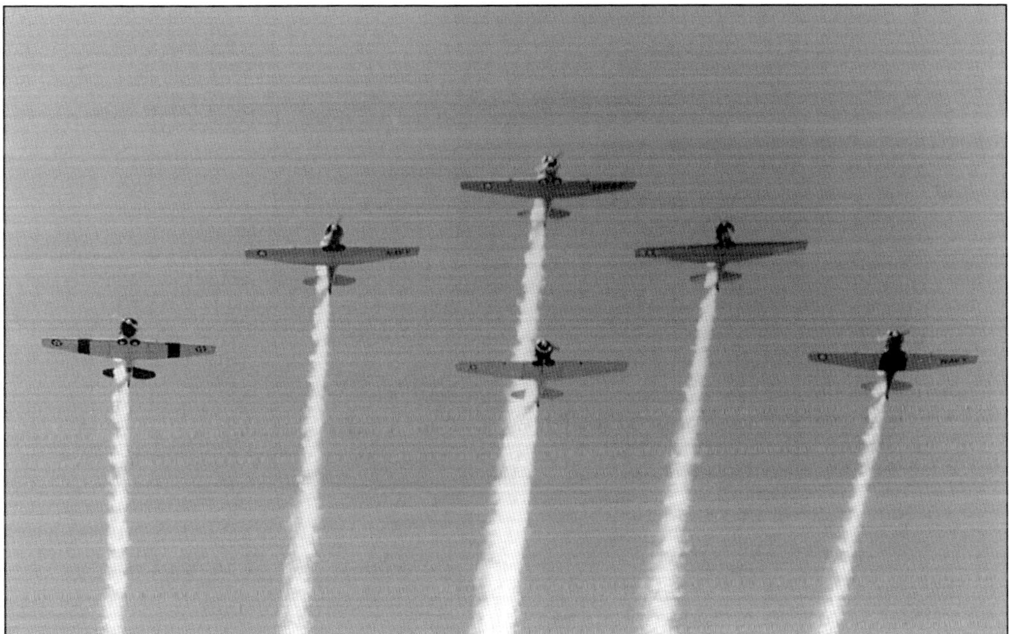

Grumman's famed fighter planes starred in World War II and 50 years later at South Weymouth air shows. The F4F Wildcat held off the Japanese during the first year of the war as a frontline fighter, joined in 1943 by the F6F Hellcat. The F8F Bearcat never made it into combat, delivered just a few weeks before the end of the war. (Navy.)

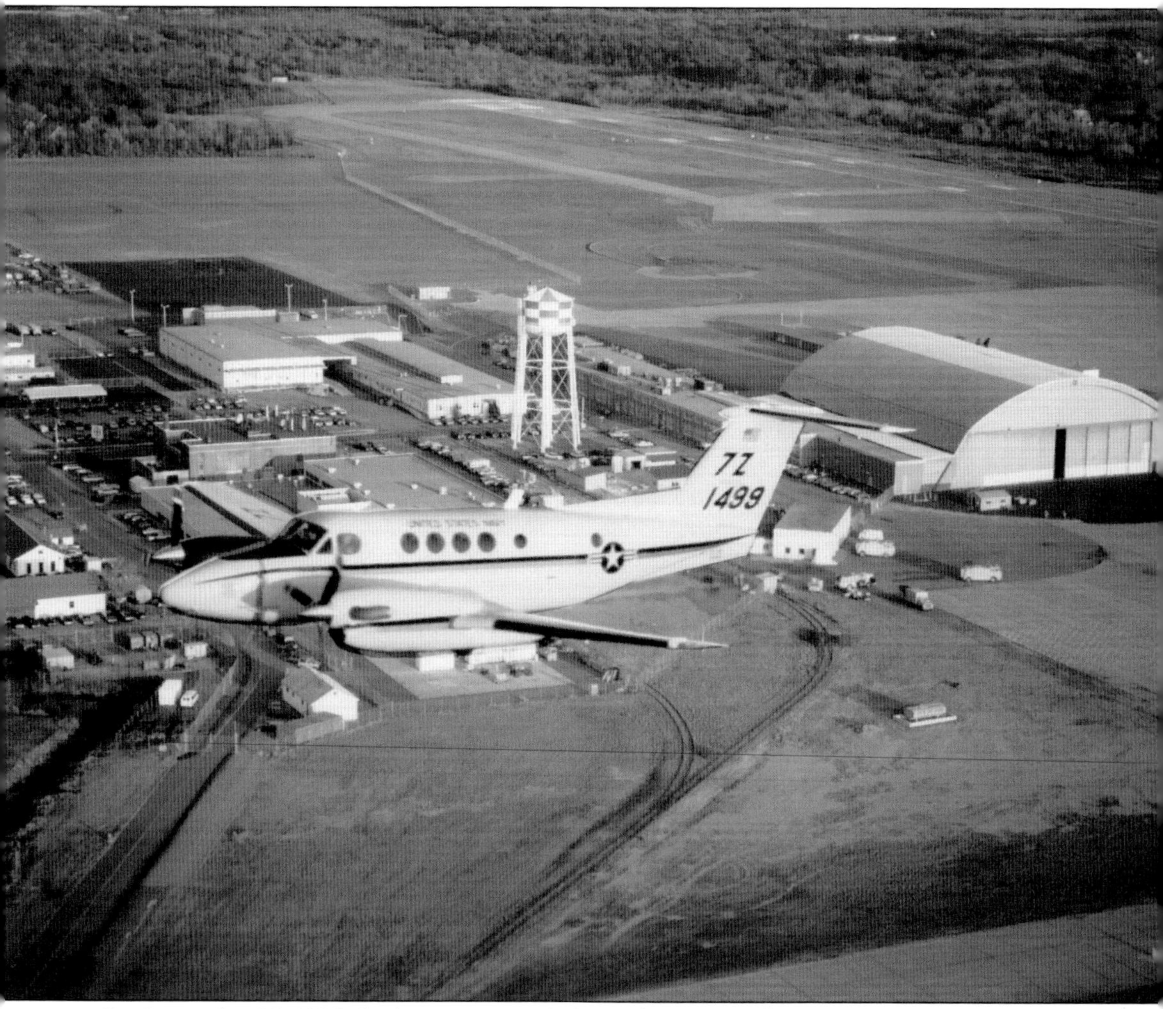

On September 30, 1996, the last operational plane, this UC-12, flew away from Naval Air Station South Weymouth, ending a longstanding, multigenerational tradition on the South Shore of Massachusetts. Machinist Larry Piller, a civilian working at the base, was one of the last people off the field. "It was an awful, awful feeling," he said in an interview. As America downsized its military and newer and better technologies made planes faster, the need for a large number of bases faded. First Squantum, then South Weymouth, fell prey to this force of history. The fact remains, though, that for about seven decades, pilots flying out of Quincy and Weymouth acted as the first line of defense against attack from the sea. Today, the Association of Naval Aviators Patriot Squadron hosts a small museum on the former South Weymouth base, ensuring that no matter what becomes of the former base, memories of the history will remain in the minds of the local citizens for a long time to come. (Navy.)

Three

THE MASSACHUSETTS AIR NATIONAL GUARD

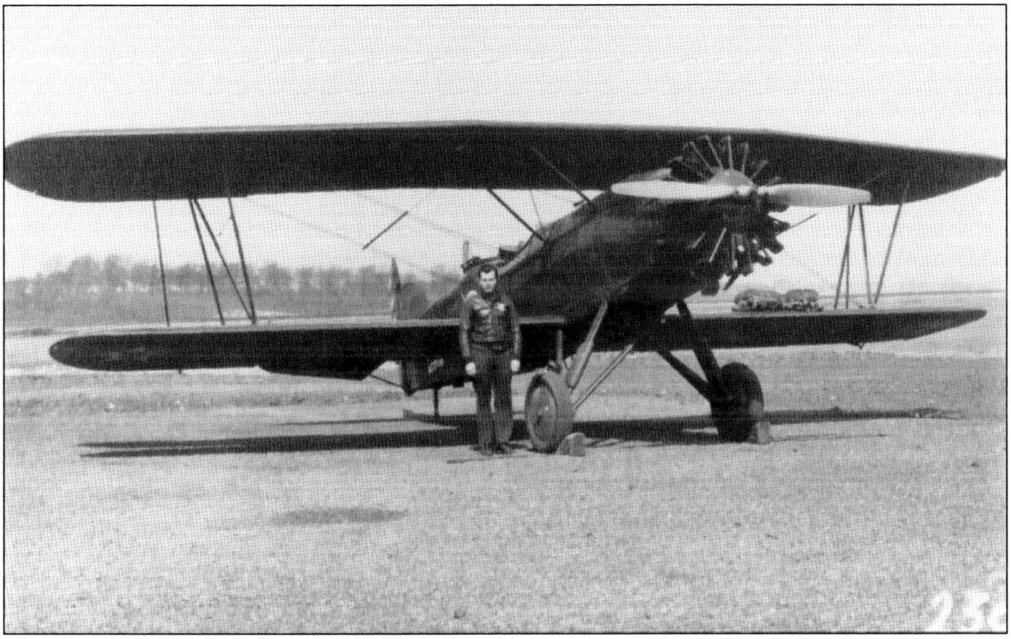

Despite successes during World War I, the potential of military aviation had barely been tapped by the Great Depression. Although proponents like Maj. Gen. Billy Mitchell touted the possibilities for planes in warfare, aviation technology did not take a great leap forward until World War II. But, thanks to feats like Charles Lindbergh's transatlantic journey in 1927, aviation caught on as a pastime for many Americans in the 1930s, leading to the construction of many small regional airports. When the Massachusetts Military Reservation was laid out between 1935 and 1940, plans included two 500-foot grass runways to be known as Otis Field. Lt. Frank Otis, a pilot and renowned Boston City Hospital surgeon, shown here with a Curtiss XO-12, died on a cross-country training mission in his Douglas O-46A on January 11, 1937, while serving with the 101st Observation Squadron. (MANGHA.)

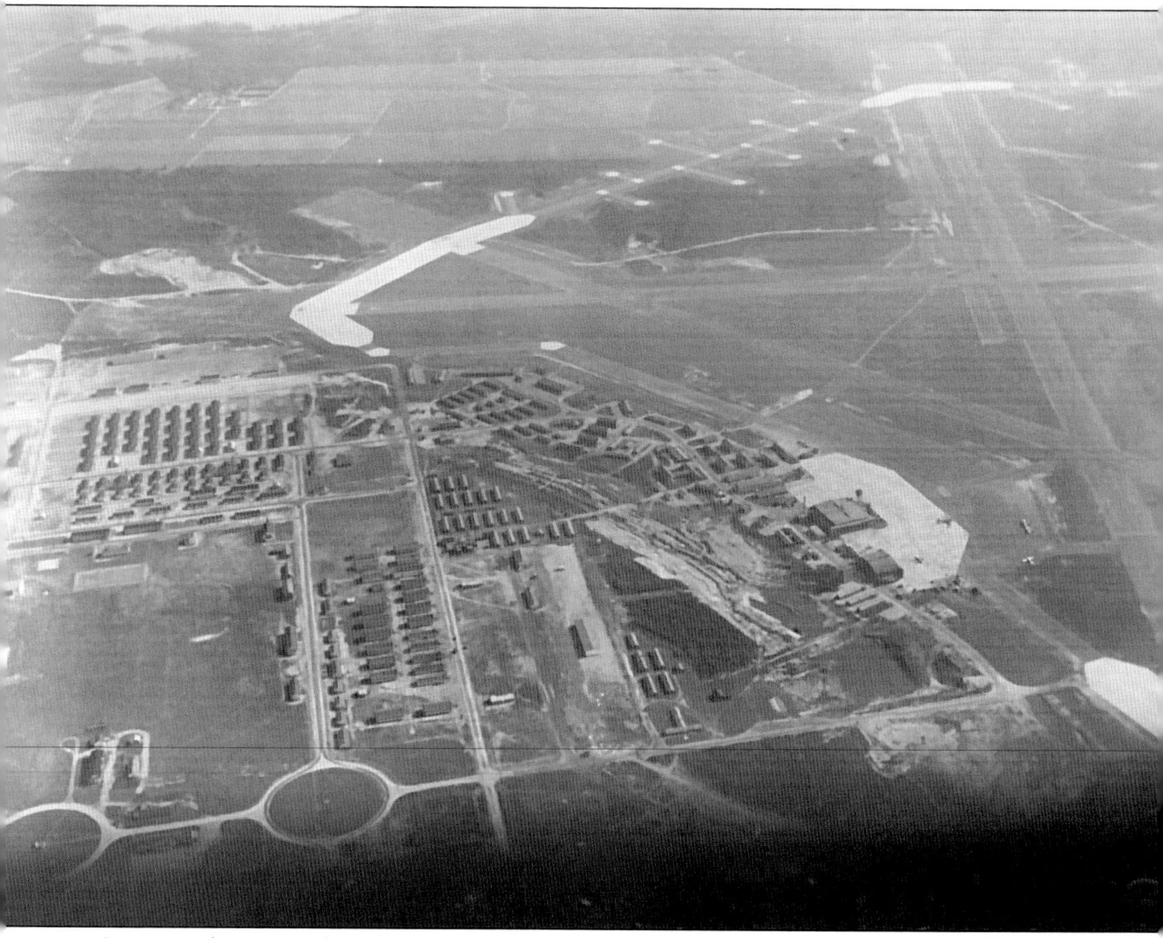

In 1940, the Massachusetts National Guard's 101st Observation Squadron, then serving at Jeffries Field in East Boston—on the site of today's Logan International Airport—was inducted into federal service and transferred to Otis Field. After 1943, the base fell under the umbrella of Naval Air Station Quonset Point, Rhode Island, being named Naval Auxiliary Air Facility Otis. Under the Navy, the base took a lead role in antisubmarine patrols of nearby waters. Photographer Walter Fleming, who shot aerials from a North American O-47, said of this photograph, "Stare at this a couple of minutes and grab a parachute. You'll probably feel as though you were falling. I did." (MANGHA.)

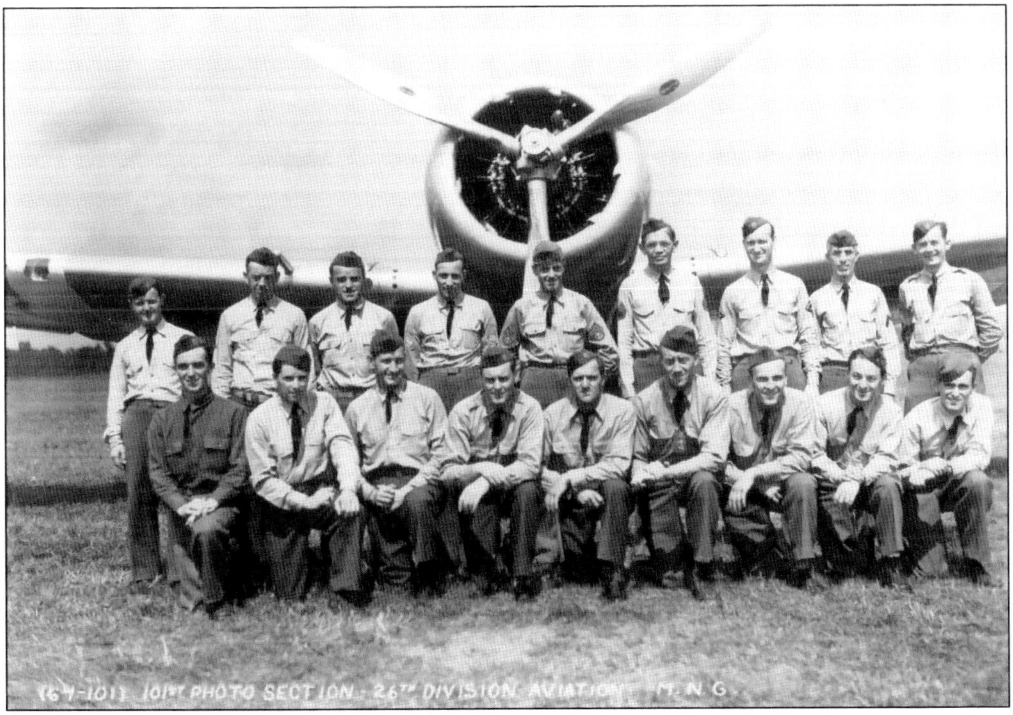

Massachusetts's 26th Infantry Division was long-famed by the time of the Great Depression. The 26th Aviation Division's photography section certainly knew how to pose a picture, and, moreover, knew how to pose for one. This one was taken on July 22, 1938. (MANGHA.)

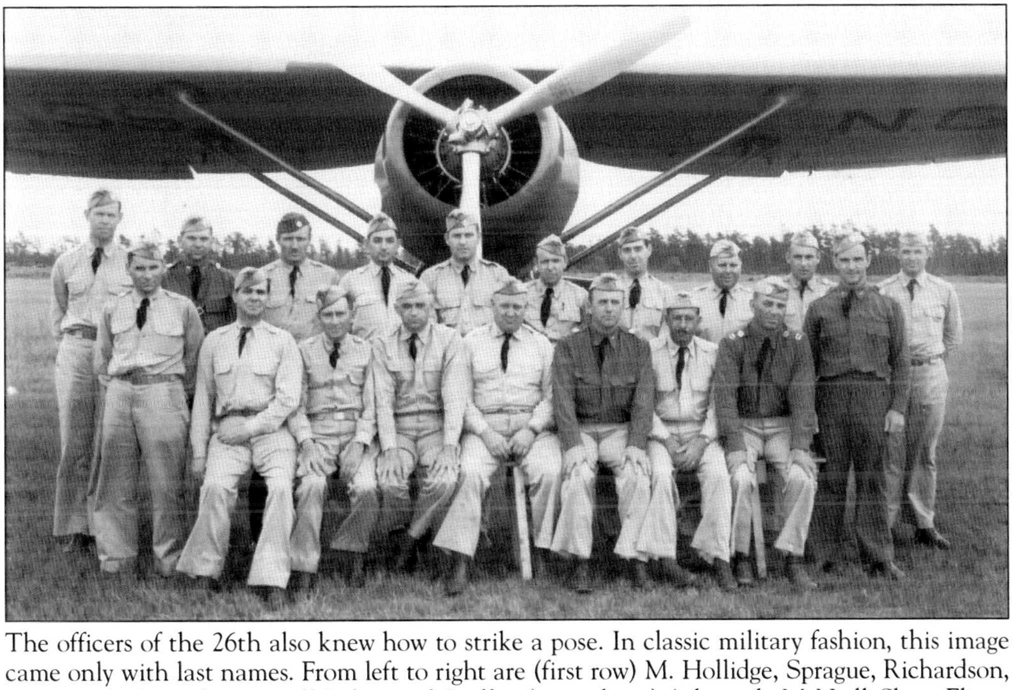
The officers of the 26th also knew how to strike a pose. In classic military fashion, this image came only with last names. From left to right are (first row) M. Hollidge, Sprague, Richardson, Stinson, Hodge, Edson, Cruff, Baker, and Griffin; (second row) Ashworth, McNeill, Shaw, Flinn, Keith, Thompson, Roullard, C. Hollidge, Jakeway, and Beck. (MANGHA.)

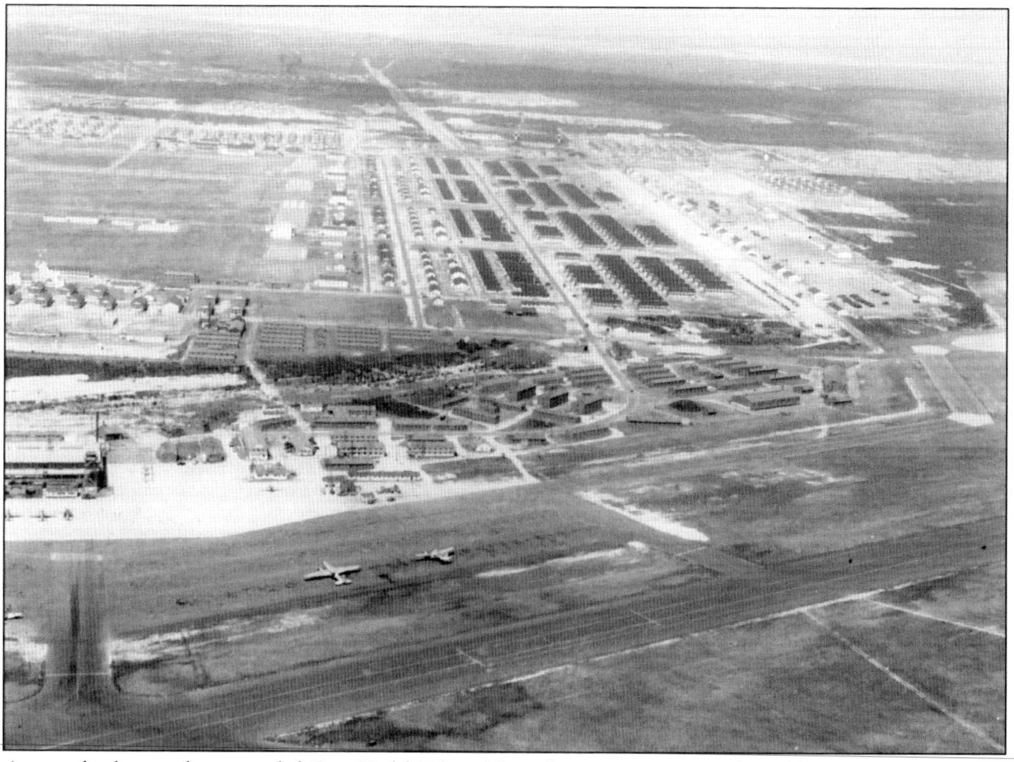

As needs changed, so too did Otis Field. The old turf runways were replaced by concrete in 1942, and the following year, those new runways were widened and lengthened. The Army Air Corps' 14th Antisubmarine Patrol Squadron utilized the field through 1943, and thereafter it fell under the Navy for the duration of World War II. Officially known as Naval Auxiliary Air Facility Otis, it served as a subordinate field under Naval Air Station Quonset Point (Rhode Island). (MANGHA.)

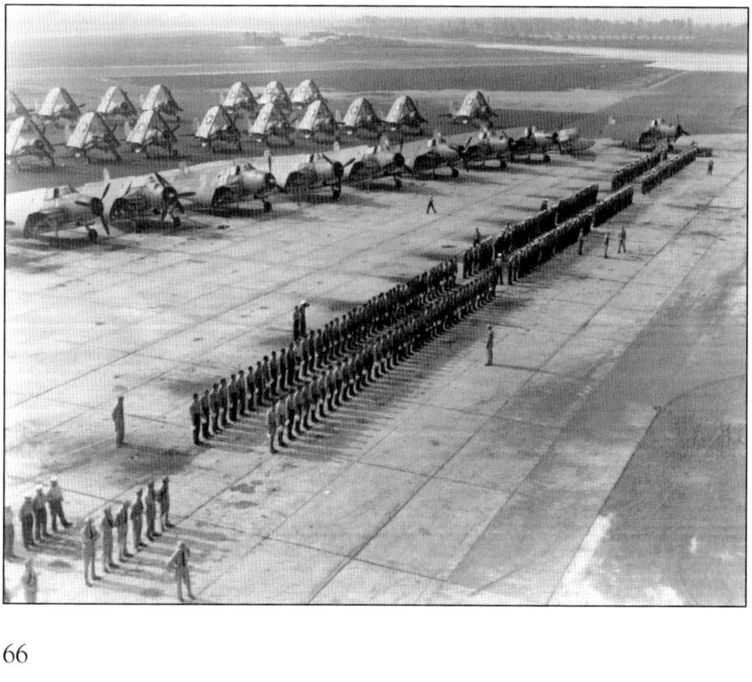

Here, an inspection is taking place at Otis Field on August 12, 1944. Four types of Navy aircraft are visible in this image: 9 TBM Avengers, 16 Vought F4U Corsairs, a Grumman J4F Widgeon, and a Lockheed PV-1 Ventura. The markings on the Avengers indicate that the planes probably belong to Carrier Group 85, soon to deploy on the USS *Shangri-La*. (Navy.)

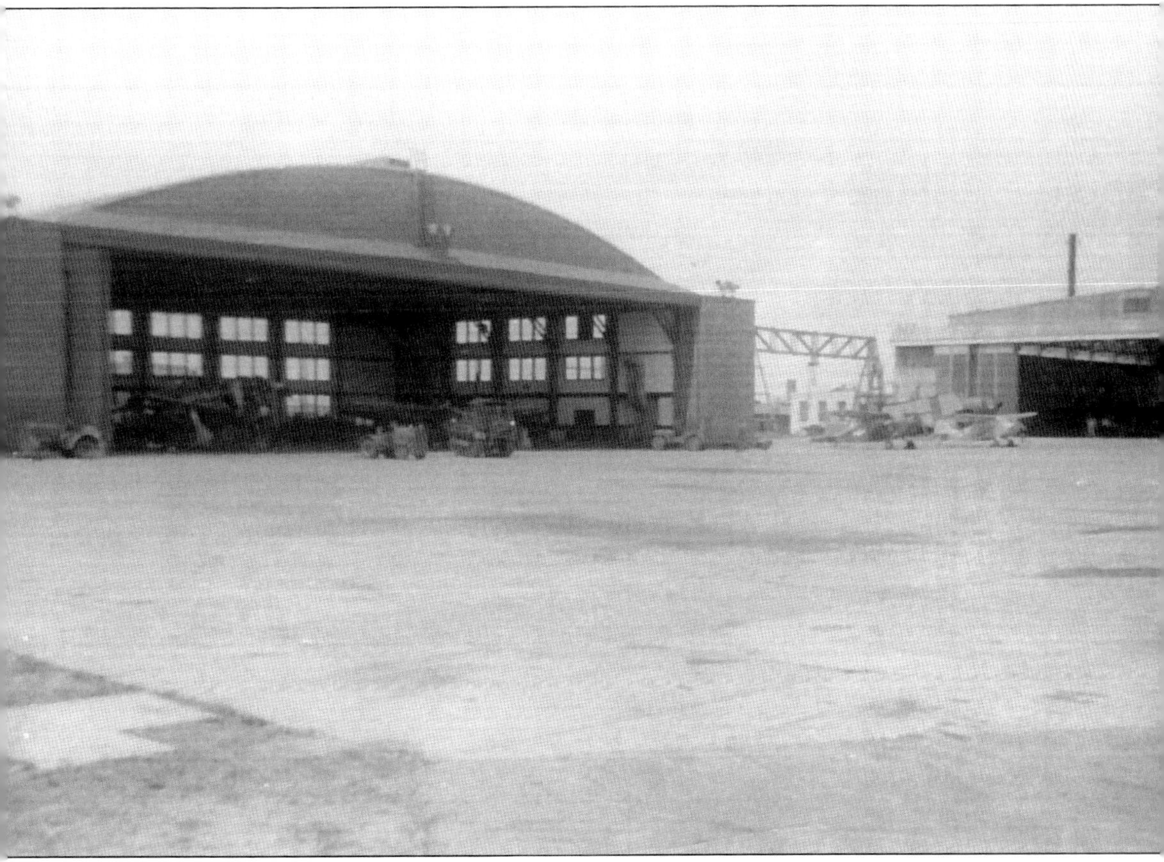

The construction of large wooden hangars for use by military aviation units was not, at first, necessary in a world already accustomed to the use of metals in construction. But, as the war advanced and the United States began rationing many items for the war effort, scrap metals were at the top of that list. Construction of military buildings with wood, like the original Otis Field hangar, served the country in two ways. (MANGHA.)

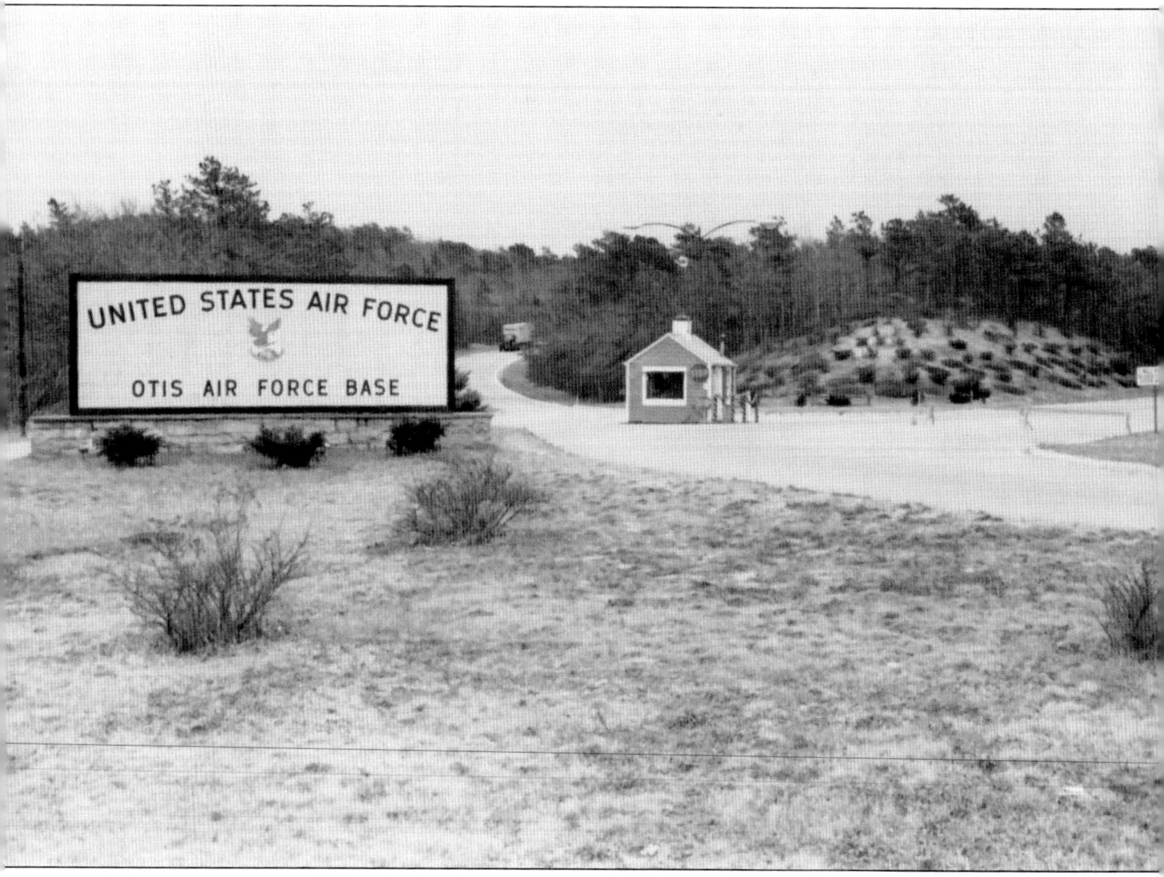

Following the end of World War II, several changes occurred that altered the military aviation story at Camp Edwards. First, in 1946, the base was deactivated and placed on caretaker status. Also in that year, the runway was lengthened to 8,000 feet, a product of the ever-growing size of American military aircraft. Then, in 1947, the Army Air Corps broke away from the US Army to become the US Air Force. When the 101st Observation Squadron was reactivated that year, the relationship between the Air Force and the National Guard was solidified. The players were all the same, just bearing different names. Different names meant it was time for new signs, like this one at the entrance to the base. In 1948, the name changed officially to Otis Air Force Base. (MANGHA.)

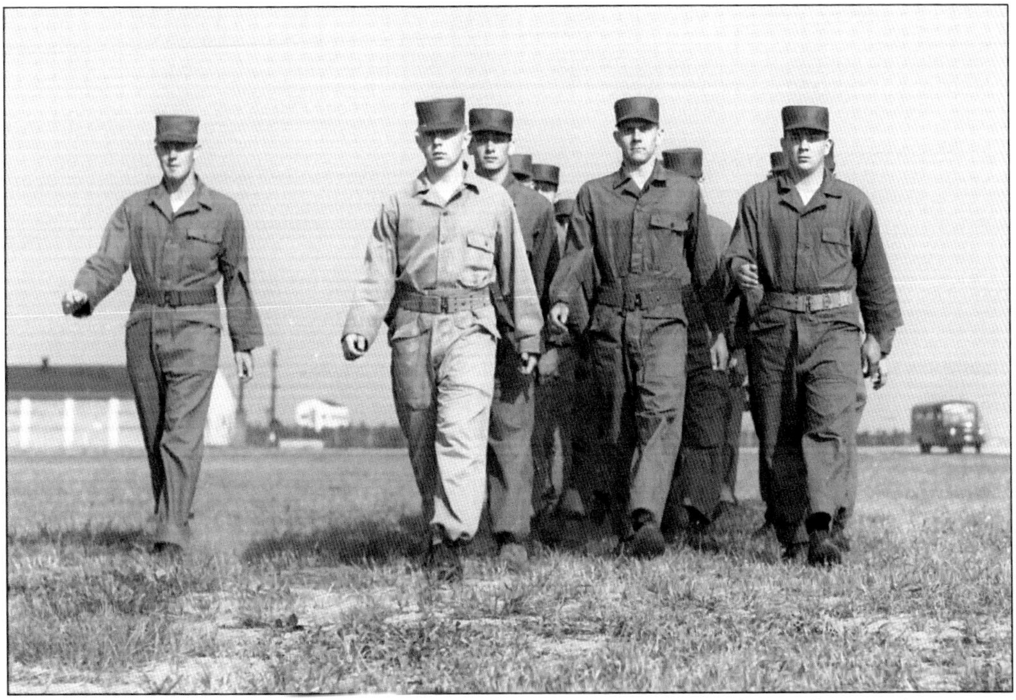

Basic military training of Air Force recruits in the 1950s became a subject of national attention with severe overcrowding of Lackland Air Force Base in Texas. Sen. Lyndon B. Johnson sent a subcommittee to investigate the situation and was told that 19,000 men were sleeping in tents at the base, for which the Air Force was remonstrated. To streamline the process, basic training was shortened from 13 to 8 weeks and new training centers opened, including one at Sampson Air Force Base in Upstate New York. The skills learned in basic training elsewhere were honed in drills and marches at duty stations, as shown here at Otis in 1955. (MANGHA.)

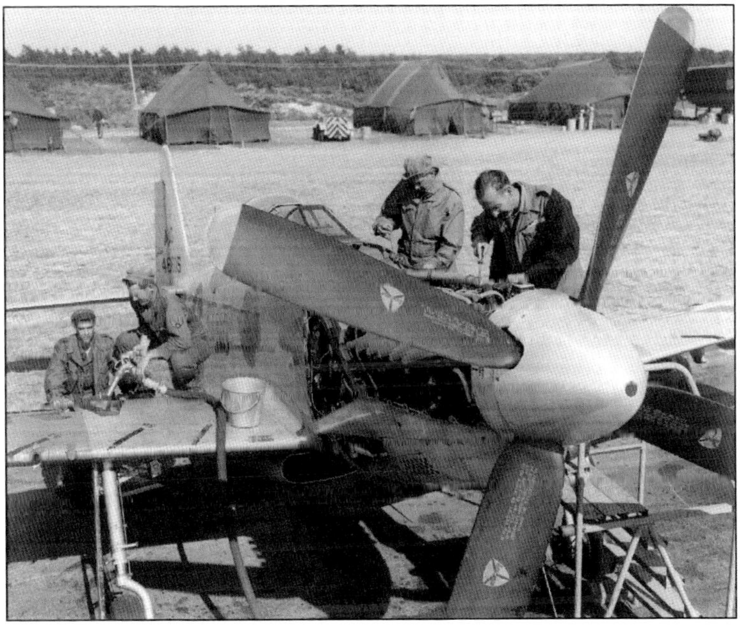

Northern pilots sought to do early season training in the south, and southern units liked to fly north, or so history would indicate. Here US Air Force Maryland Air National Guard mechanics work on a North American F-51H-10-NA Mustang assigned to the 104th Fighter-Bomber Squadron during the 1954 Summer Encampment at Otis Air Force Base. The Maryland National Guard flew F-51s from 1951 to 1955. (USAF.)

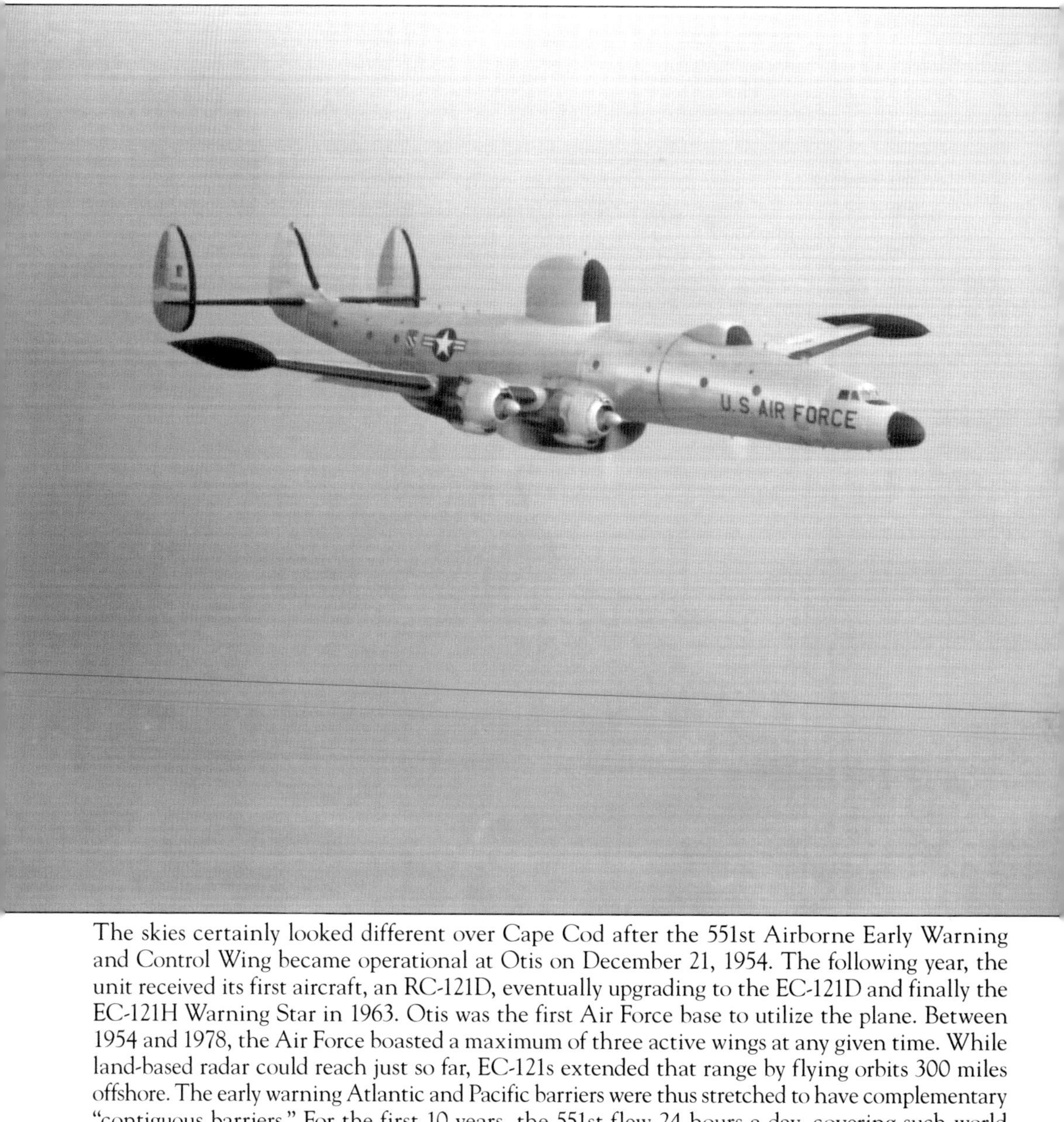

The skies certainly looked different over Cape Cod after the 551st Airborne Early Warning and Control Wing became operational at Otis on December 21, 1954. The following year, the unit received its first aircraft, an RC-121D, eventually upgrading to the EC-121D and finally the EC-121H Warning Star in 1963. Otis was the first Air Force base to utilize the plane. Between 1954 and 1978, the Air Force boasted a maximum of three active wings at any given time. While land-based radar could reach just so far, EC-121s extended that range by flying orbits 300 miles offshore. The early warning Atlantic and Pacific barriers were thus stretched to have complementary "contiguous barriers." For the first 10 years, the 551st flew 24 hours a day, covering such world events as overflights of Cuba during the Cuban Missile Crisis in 1962. The wing was deactivated on the last day of 1969. (MANGHA.)

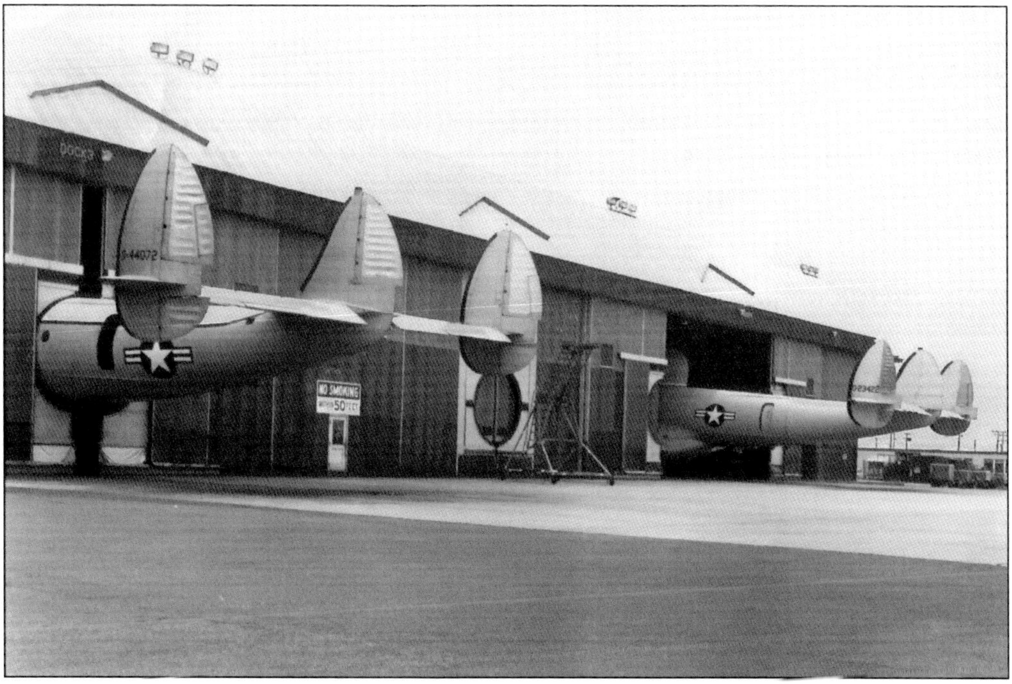

Not all was ready for the arrival of the Warning Stars. Without hangars long enough to accommodate the entirety of the 113-foot-long planes, the 551st was forced to retrofit doors to close around the tail sections of the craft, leaving them exposed to the elements. (MANGHA.)

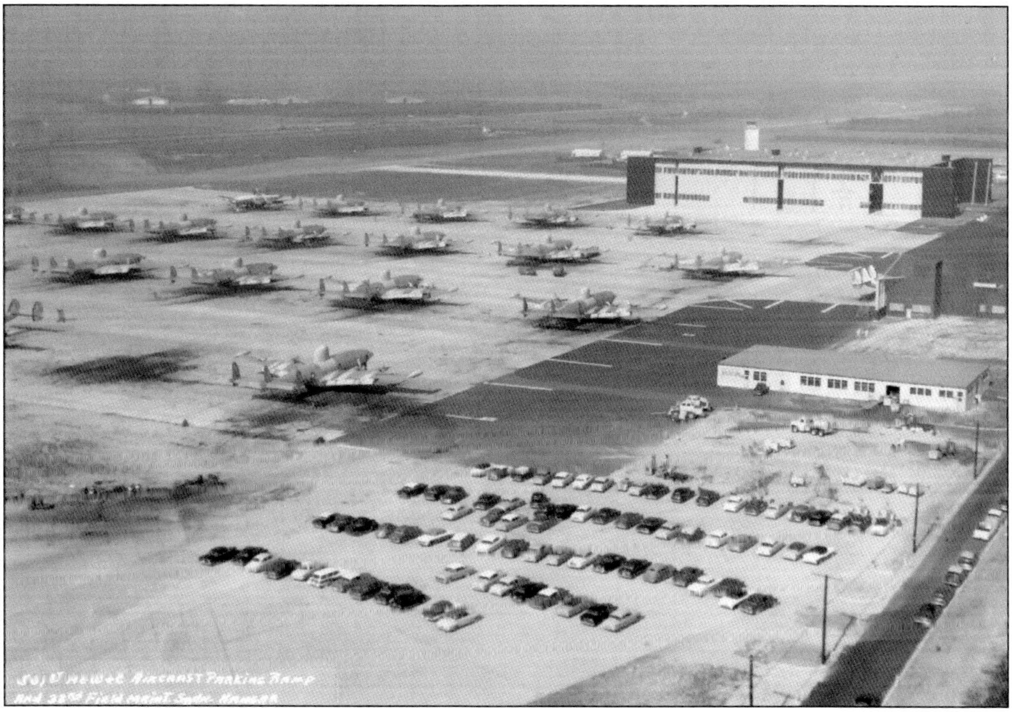

According to the history of the 102nd Fighter Wing, "Until 1973, Otis was the largest Aerospace Defense Command base in the world and the only Air Force base named for a doctor." (MNG.)

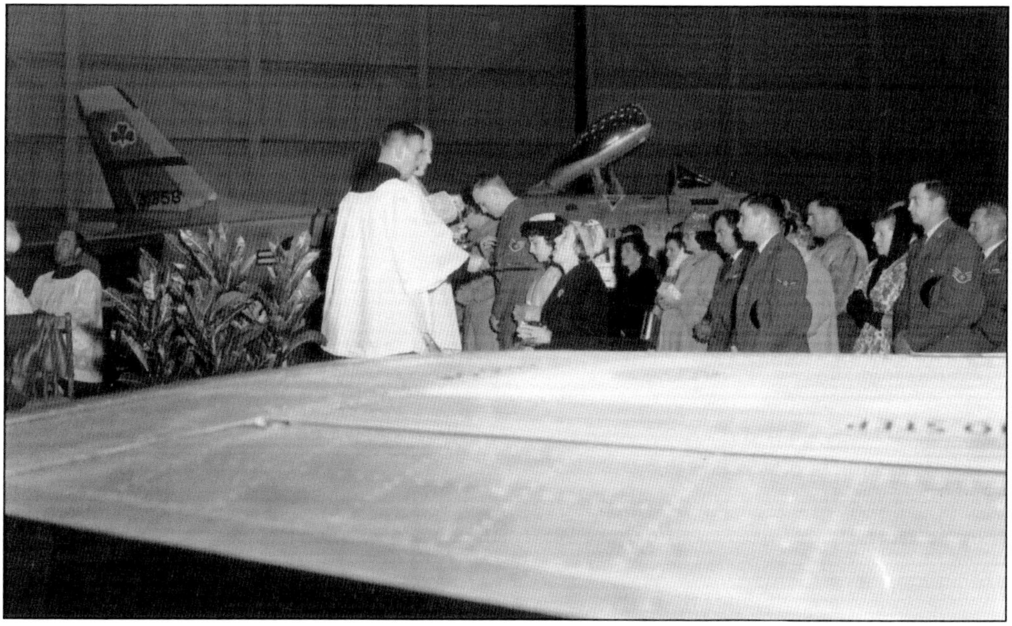

Cold War tensions reached their heights in the early 1960s with the Cuban Missile Crisis, the U-2 incident, and the Berlin Crisis. As the Soviets threatened to close West Berlin to the West, Pres. John F. Kennedy called for an escalation in troop strength and an increase in military spending. By October, the Air National Guard's 102nd Tactical Fighter Wing was recalled and prepared to go overseas. Cardinal Richard Cushing distributed communion to airmen just before their departure to Phalsbourg, France, in late 1961. (MANGHA.)

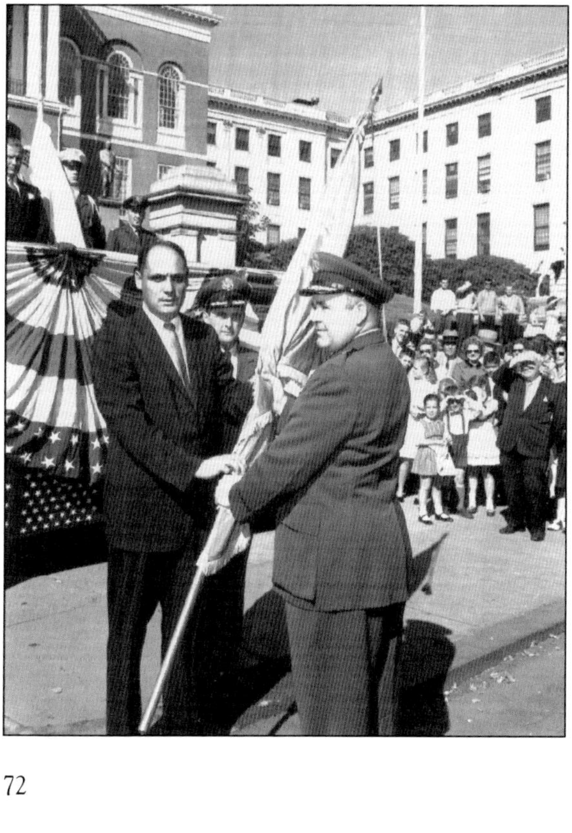

With 82 F-86 Sabres, the 102nd crossed the Atlantic with a mission of close air support for NATO ground forces should any military action arise, as well as interdiction of enemy aircraft. The men of the 102nd participated in joint exercises with foreign troops, but political tensions eased enough that the unit was called home on May 7, 1962. Brig. Gen. Charles W. Sweeney, shown here, led the 102nd there and back. (MANGHA.)

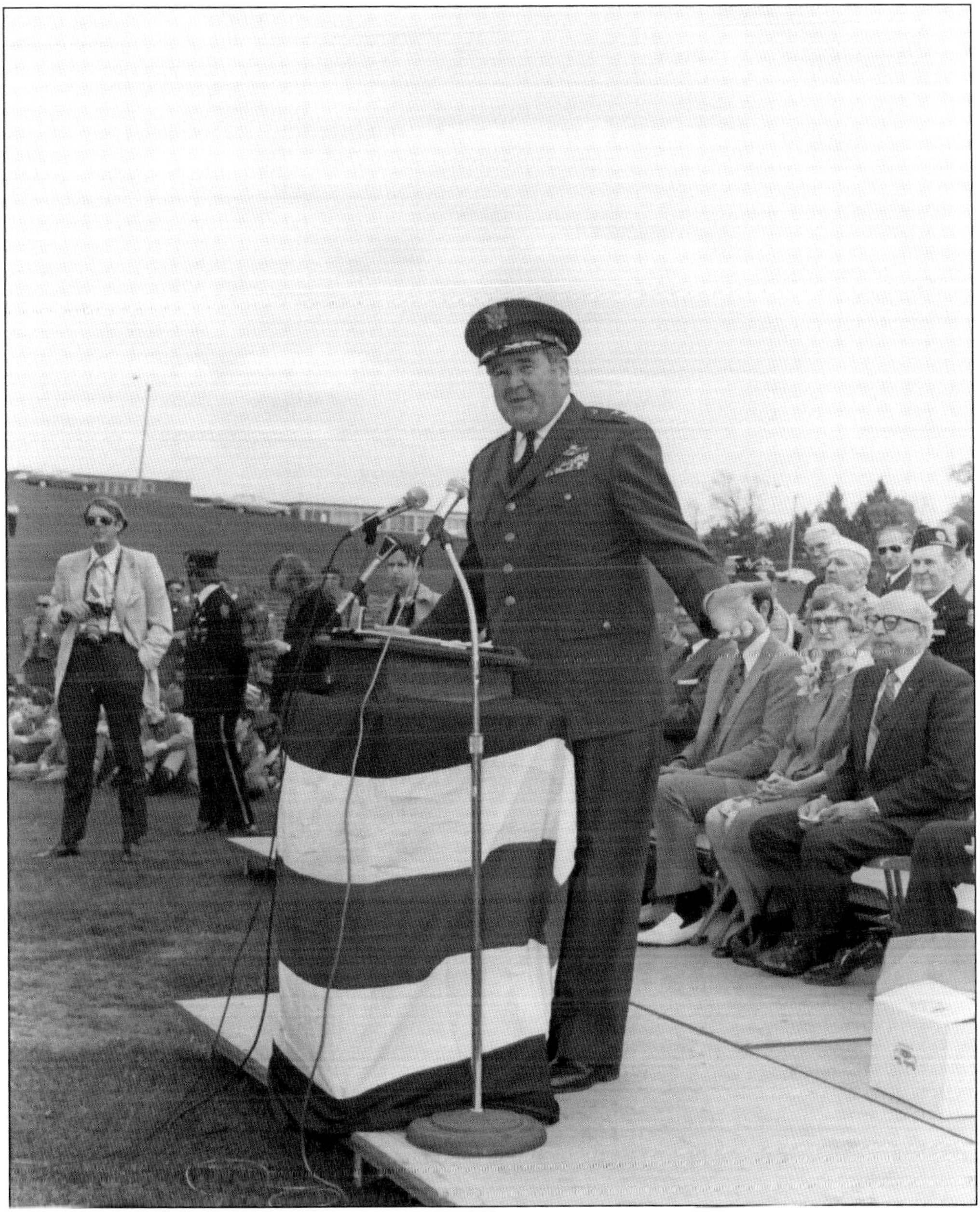

General Sweeney was a story in himself. Born in Lowell on December 27, 1919, he learned to fly in Quincy while in high school, joining the US Army in 1941 as an air cadet. He spent the war in preparation for its last offensive. As commander of the 393rd Bombardment Squadron in July 1945, he flew test flights to ready himself for the first bomb he would drop on the enemy after four years in service. On August 9, 1945, he commanded *Bockscar*, a B-29 Superfortress, headed for the Japanese city of Kokura, but weather diverted the flight, sending the plane in search of a secondary target. Over Nagasaki, *Bockscar* dropped Fat Man, a plutonium bomb that killed at least 70,000 Japanese and led to the end of World War II six days later. From 1955 to 1976, General Sweeney commanded the Massachusetts Air National Guard's 102nd Tactical Fighter Wing. (MANGHA.)

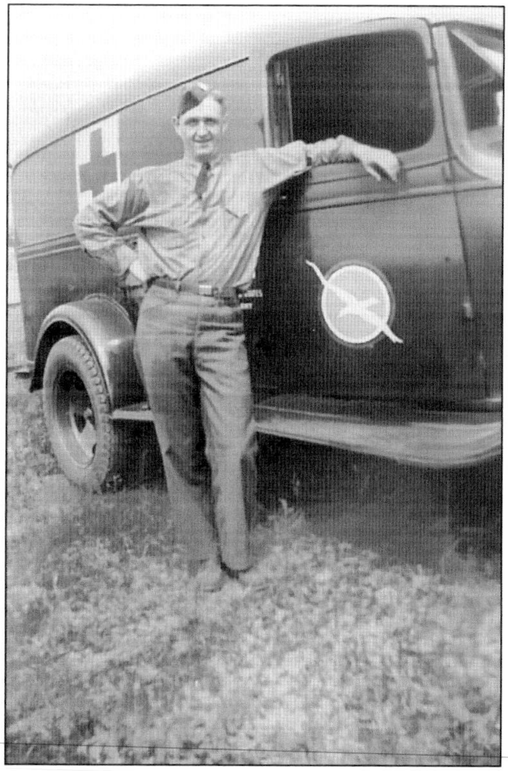

Carl Allen showcases the insignia of the Massachusetts Air National Guard in this 1939 photograph, designed by noted artist Paul F. Seavey in 1924. The blue field once stood for the waters of Boston Harbor, and today it stands for the skies above Massachusetts and the Atlantic Ocean surrounding Cape Cod. The gold ring represents unity, and the gull symbolizes vigilance and keen vision. (MANGHA.)

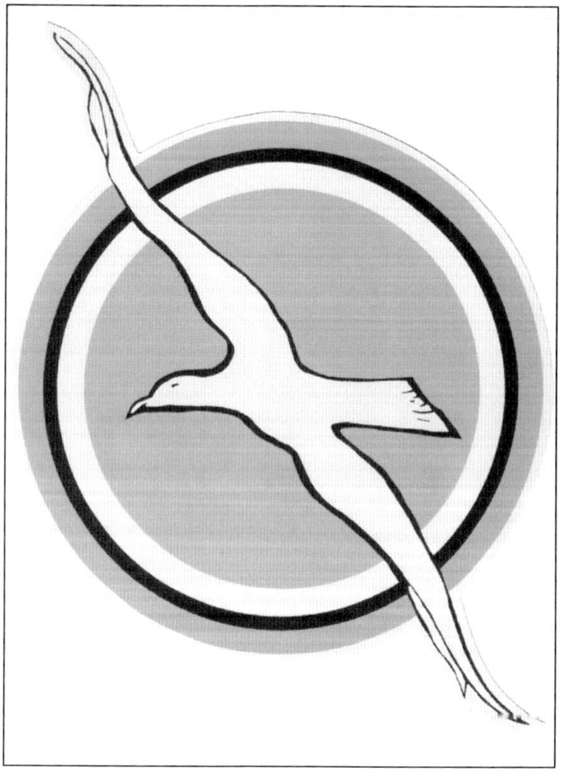

The Massachusetts Air National Guard changed from the gull design to the Minuteman with time. Today, the state's military air wing operates three units, the 102nd Intelligence Wing and the 253rd Cyberspace Engineering Installation Group at Otis and the 104th Fighter Wing out of Barnes Air National Guard Base in Westfield. (MANGHA.)

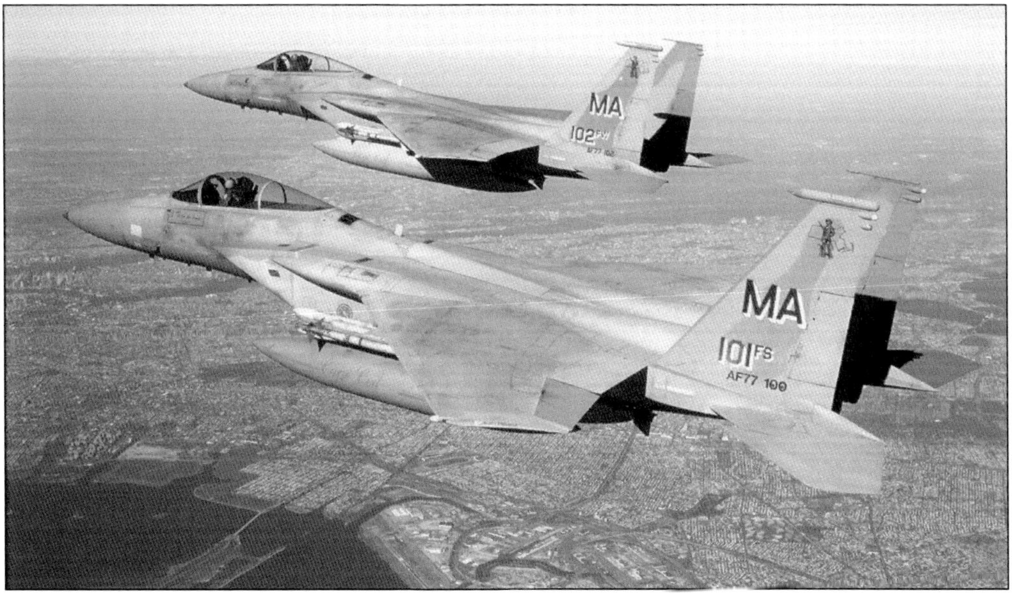

In 1973, the Air Force deactivated Otis Air Force Base, meaning only that the national units left the base and that the task of 24-hour-a-day surveillance fell to the 102nd Fighter Wing of the Massachusetts Air National Guard. The Massachusetts Army National Guard took command of the rest of Camp Edwards, with the Coast Guard utilizing a portion of the base. The 102nd moved from F-84F Thunderstreaks to F-100D Supersabres to F-106 Delta Darts to F-15 Eagles, shown here. (MNG.)

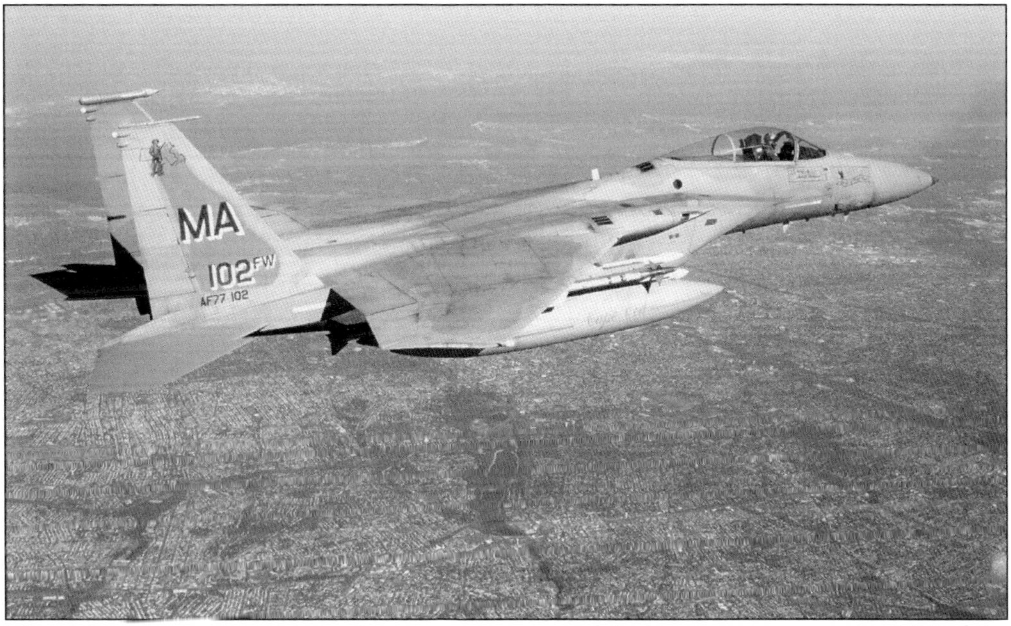

Through various interception and training missions and deployments overseas, the F-15s of the 102nd bore the Minuteman insignia of the Massachusetts National Guard on the tail and the gull insignia just aft and below the cockpit. In 1999 and 2000, the unit deployed to the Middle East and Europe to take part in Operations Northern Watch and Southern Watch, enforcing no-fly zones around Iraq. (MNG.)

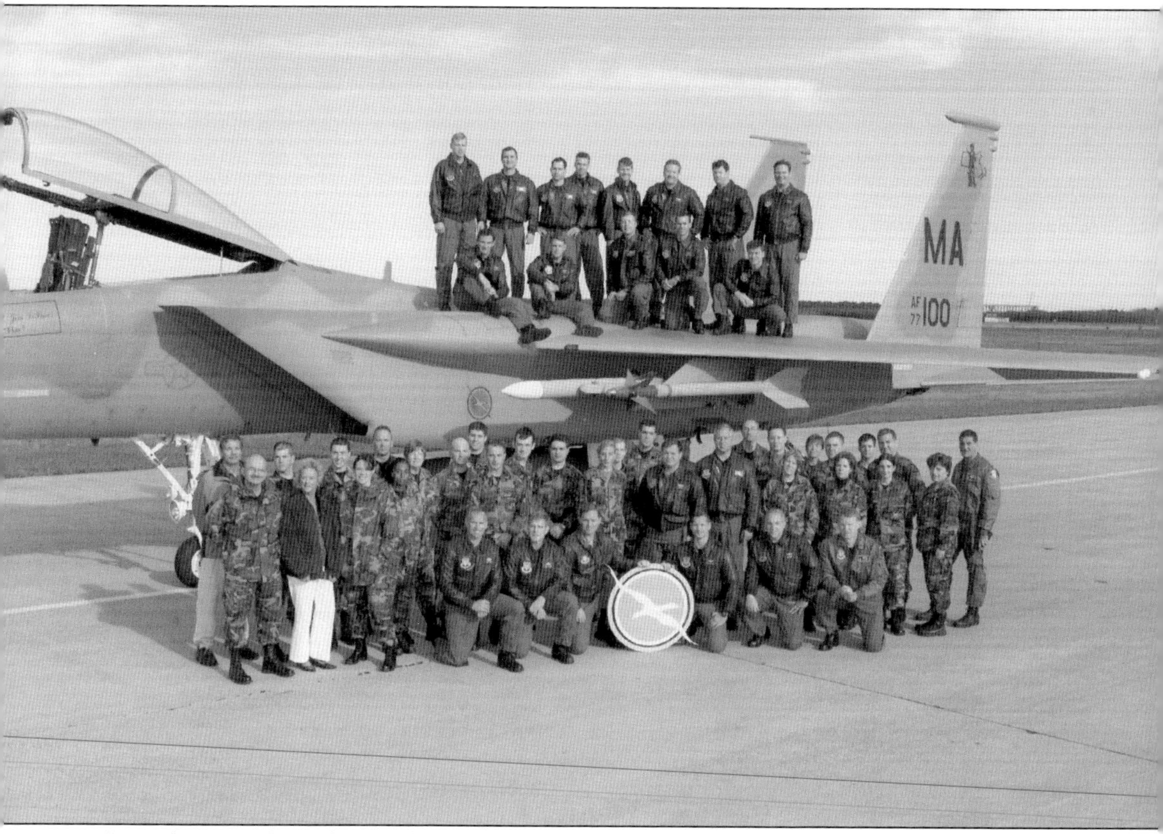

According to Col. Anthony Schiavi, commander of the 102nd Intelligence Wing, on September 11, 2001, Lt. Col. Tim Duffy and Maj. Dan Nash were the first responders to the terrorist attacks on the World Trade Center towers in Manhattan. They arrived just moments after the second strike "and would be the sole ID authority for aircraft, helicopters, etc. that were still airborne after the towers were hit, ensuring that there were no additional threats to the city." For the next six months, the 102nd Fighter Wing flew nonstop patrols as part of Operation Noble Eagle. A closure scare in 2005 threatened the end of the 102nd, but it instead made the transition to an intelligence wing on April 1, 2008, to aid in the War on Terror, losing its planes but maintaining its operational status. As of this writing, the Camp Edwards and Otis Air Base stories continue. (MNG.)

Four

The Coast Guard

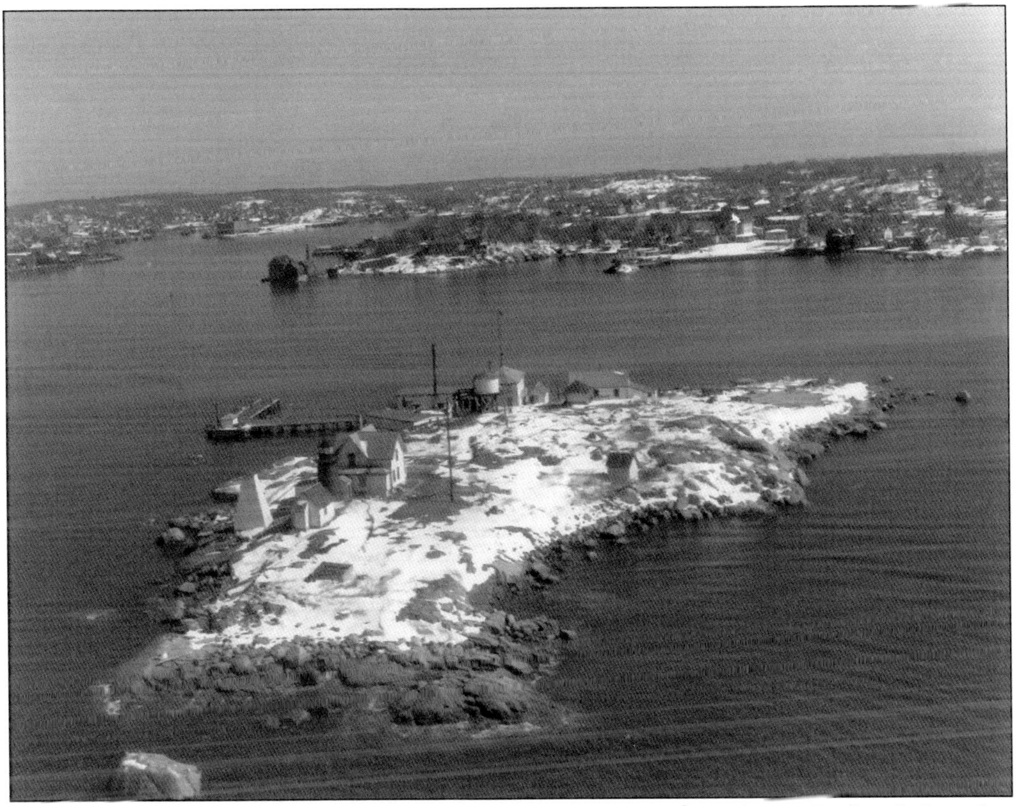

Prohibition inspired the Coast Guard to take to the skies. Comdr. C.C. von Paulsen believed in the power of aerial searches in the service's appointed task of stopping rumrunners and as such sought out a surplus Navy Vought UO-1 seaplane being held at Cape May, New Jersey. He had it flown out of the Squantum Naval Air Station before transferring it to a dormant US Fisheries facility called Ten Pound Island off Gloucester in 1925, where he had charge of Coast Guard Station No. 7. (USCG.)

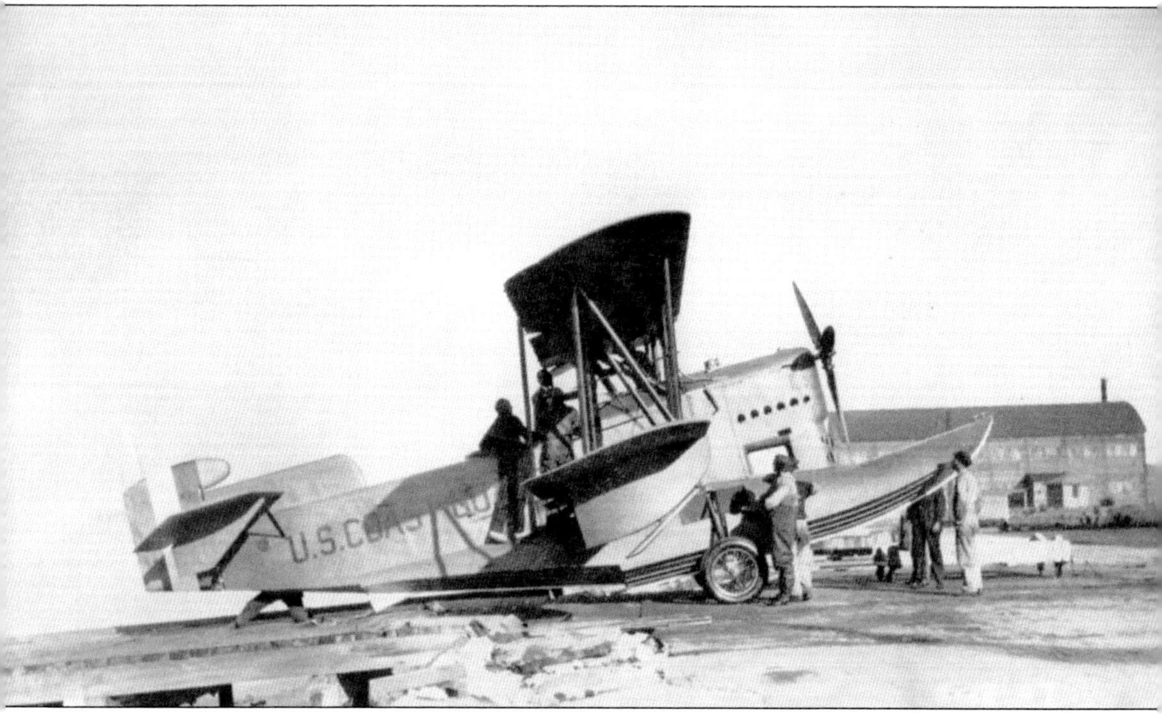

In October 1926, von Paulsen and his crew at Gloucester took delivery of three of the Coast Guard's first five planes, in their case Loening OL-5 amphibians. Congress had recently appropriated $152,000 to develop a permanent Coast Guard aviation detachment. As these planes were specially designed for Coast Guard use, and as the service had previously used only surplus Navy planes, the Coast Guard pinpoints this moment as the true beginning of Coast Guard aviation. Von Paulsen said of the OL-5, speaking to Grover Loening for a 1970s history of the airplane manufacturer, "That airplane of yours was a real departure at that time. It was a leader—a good, sound plane with a sound engine that gave no trouble if properly cared for. As for the plane—it flew so easily that I never gave its 'deadbeat' flying characteristics much thought." (USCG.)

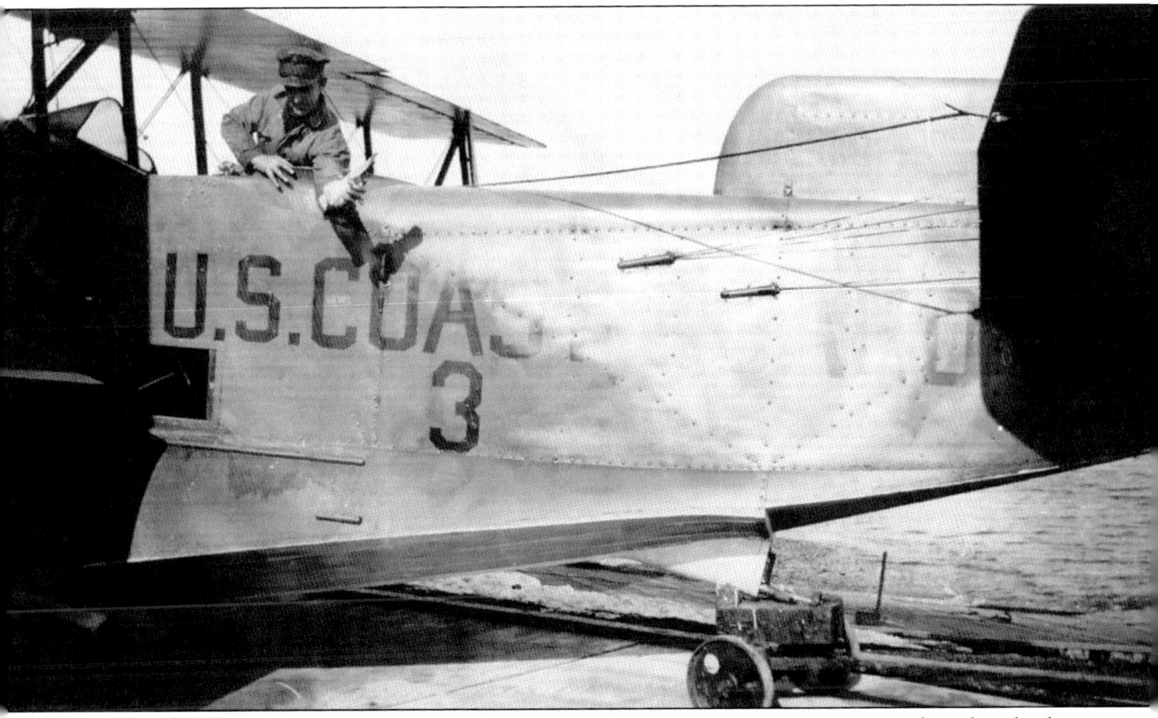

The Coast Guard's Loenings differed from their Navy and Army counterparts in that they had increased fuel capacity and were painted with a product specifically developed for the plane itself, a "high-visibility-over-water-chrome yellow," according to the Coast Guard Historian's Office. The manufacturer strongly reinforced the hull bottoms and keels, and special skids protected the wing-tip floats. For stability and maneuverability, the planes had extra fins and rudders placed higher than usual. Operationally, the planes broke new ground by landing on water, effectively opening up new avenues for rescue at sea. As a product of the times, during the height of Prohibition, Loening pilots experimented with machine guns and eventually battery-powered radios. Here, though, a pilot demonstrates the use of pigeons in air-to-ground communication as one of the Loenings leaves the base to meet Charles Lindbergh on May 18, 1927. (USCG.)

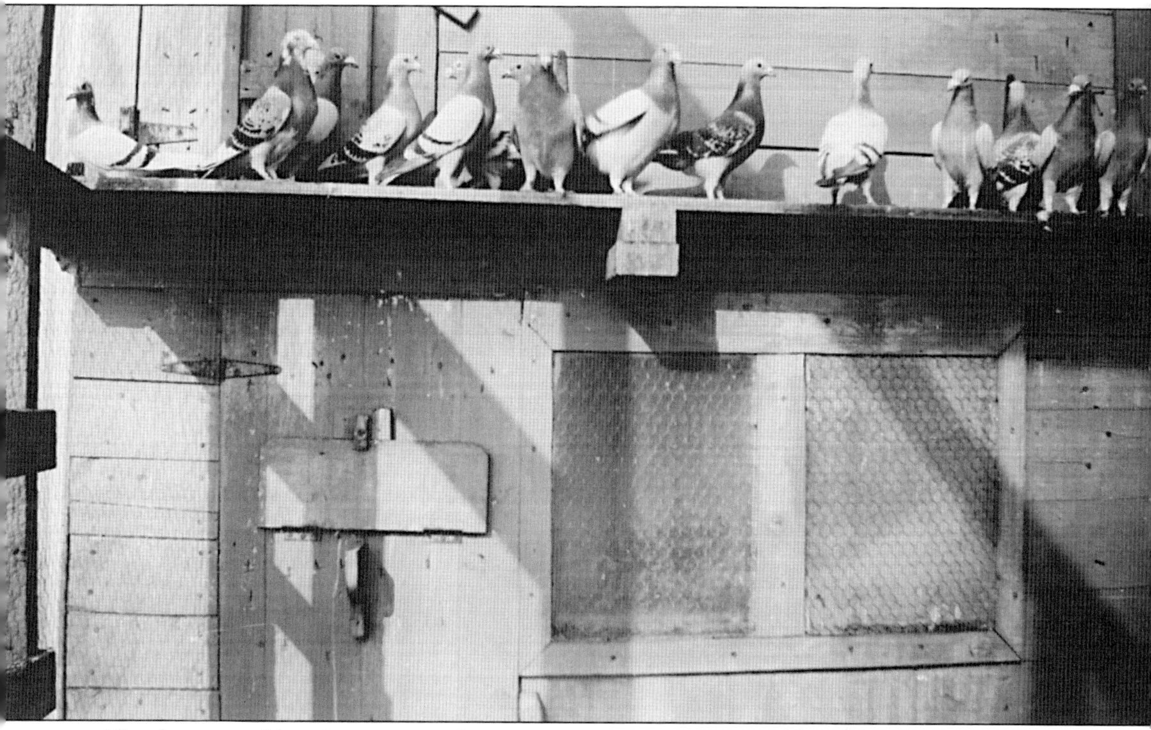

The first use of homing pigeons for carrying messages is traceable to the ancient Olympiads 3,000 years ago, and the practice has been repeatedly experimented with by different civilizations through time with varying degrees of success. Lighthouse keepers used them in the 1800s, and spy networks would utilize them in World War II, but in between, it was the Coast Guard who found the most interesting application of the birds' ability to find their way home from any release point. An article in *Popular Mechanics* in October 29 detailed how the Coast Guard flyers from Ten Pound Island used them to spot schools of fish for the local fishing fleet on and around Stellwagen Bank. After finding a large school of mackerel, the pilot tucked a note into the tube strapped to his pigeon's leg. "With wings folded tight, the homer dives straight down to the sea, almost within reach of the spray of the waves. He circles three or four times and then darts toward the pigeoncote on Ten Pound Island, at Gloucester." (USCG.)

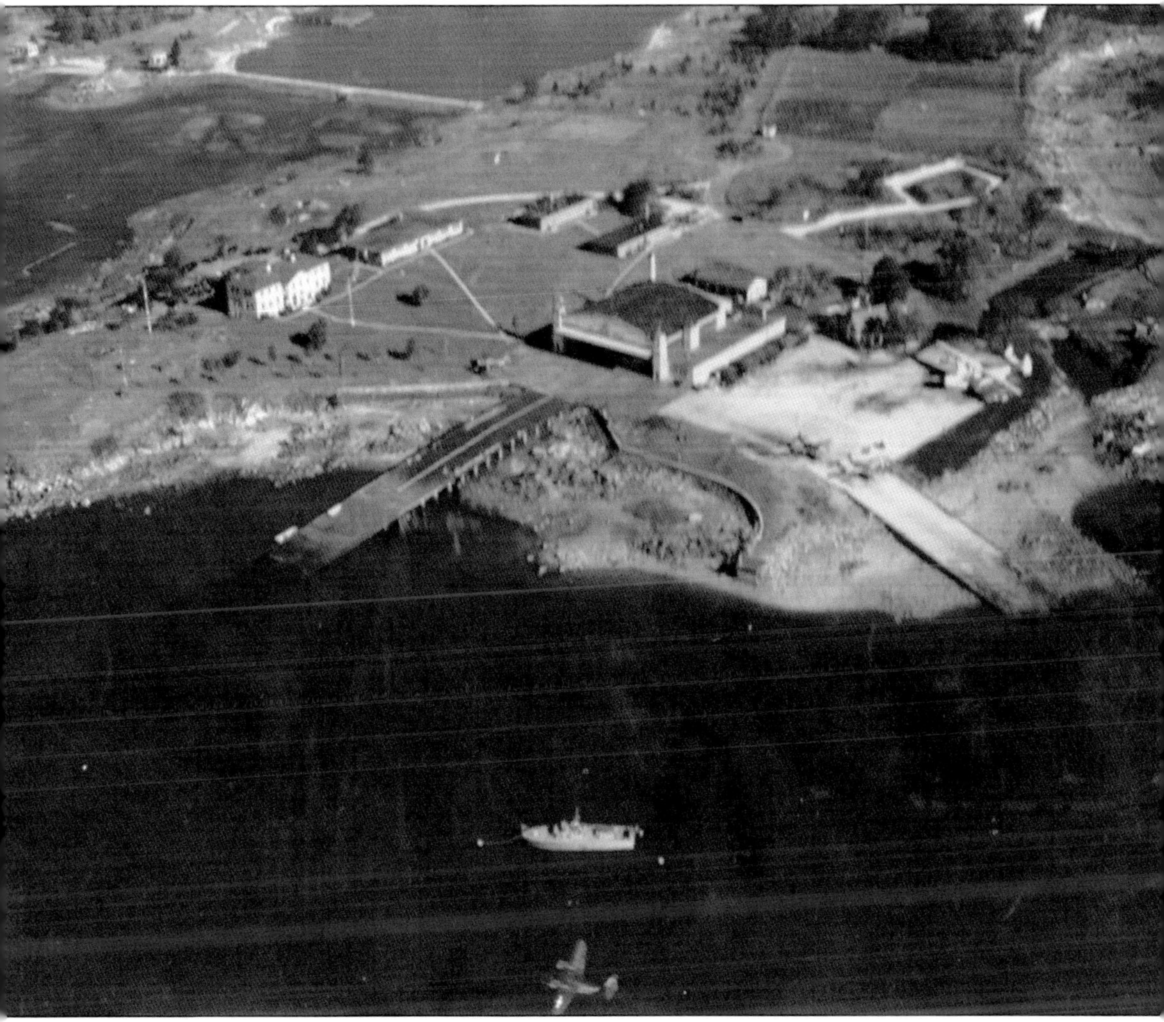

Due to the success of the operations at Ten Pound Island and, indeed, the overall popularity of aviation in general after Lucky Lindbergh's 1927 transatlantic flight, the Coast Guard soon outgrew its Gloucester home. Knowing that aviation had become a permanent component of the service's many missions (antismuggling, iceberg tracking, search and rescue, and more), the Coast Guard sought a new base on the North Shore. It found it on Winter Island in Salem. The new base featured a hangar (Ten Pound Island has started with a surplus US Army tent), two runways down which the seaplanes would run between dry land and the sea, and a paved parking apron. It also housed many state-of-the-art facilities that helped prepare pilots and properly equip the planes for flight. Air Station Salem opened on February 15, 1935, and during its first year recorded 26 medical evacuation cases. As at Gloucester, though, many other missions crept into the normal day's work. (USCG.)

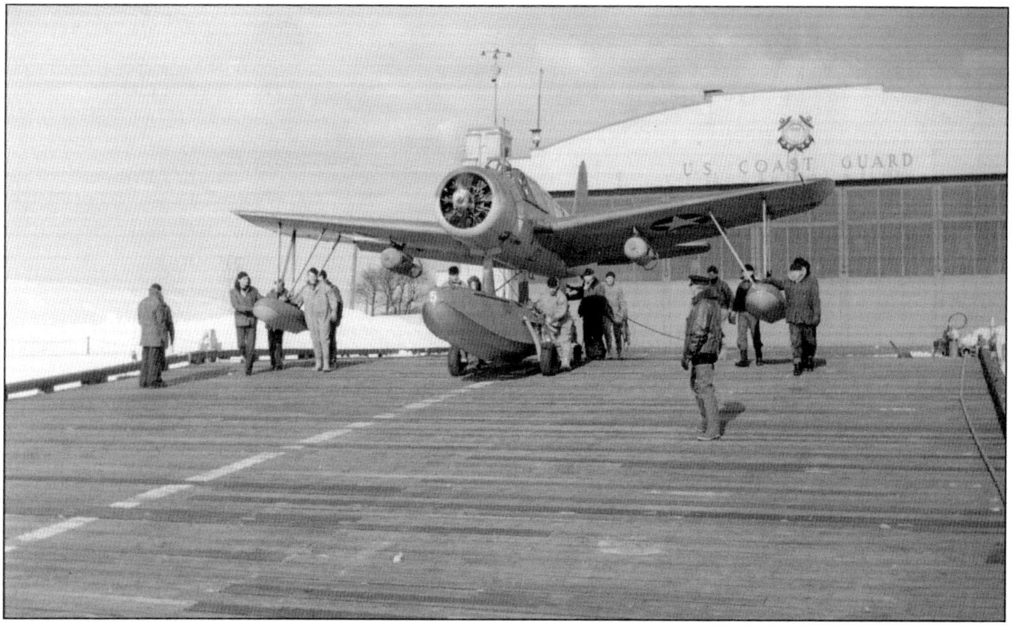

The Vought OS2U Kingfisher stood among the many varieties of planes used at Salem during its operational lifetime, from 1935 to 1970. At its height during World War II, the station rotated 37 planes through its mission cycle. Missions included antisubmarine patrols, search and rescue, and even counting migratory waterfowl for the US Biological Survey. Rumrunning was no longer an issue, as Prohibition ended in 1933. (USCG.)

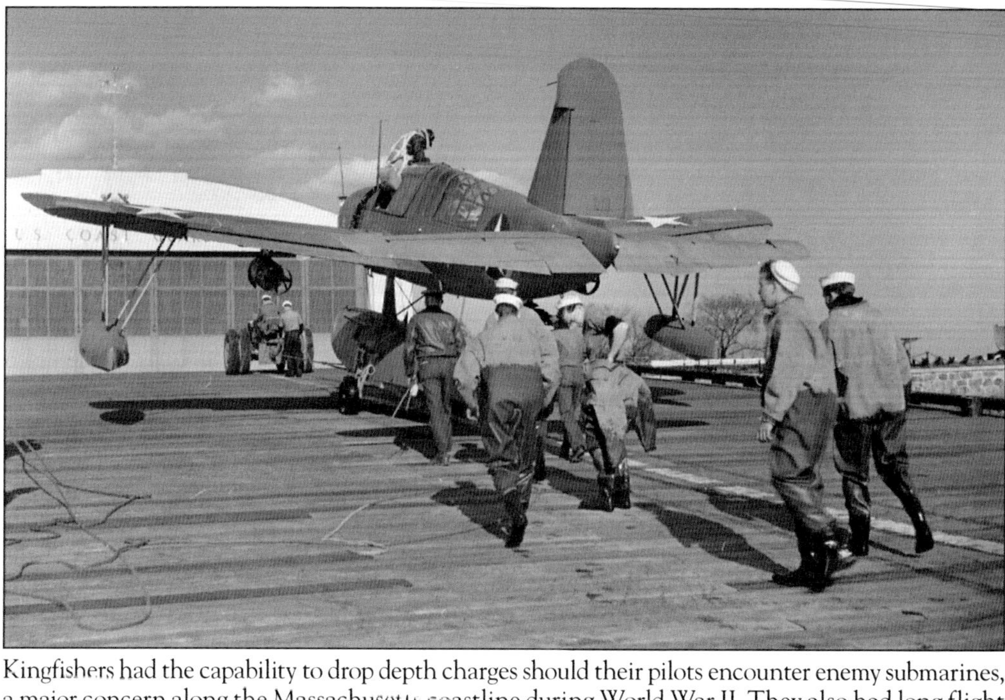

Kingfishers had the capability to drop depth charges should their pilots encounter enemy submarines, a major concern along the Massachusetts coastline during World War II. They also had long flight capabilities for their time, as much as six hours per mission. Yet, with all that, no Coast Guard crew flying a Kingfisher ever scored a U-boat kill. (USCG.)

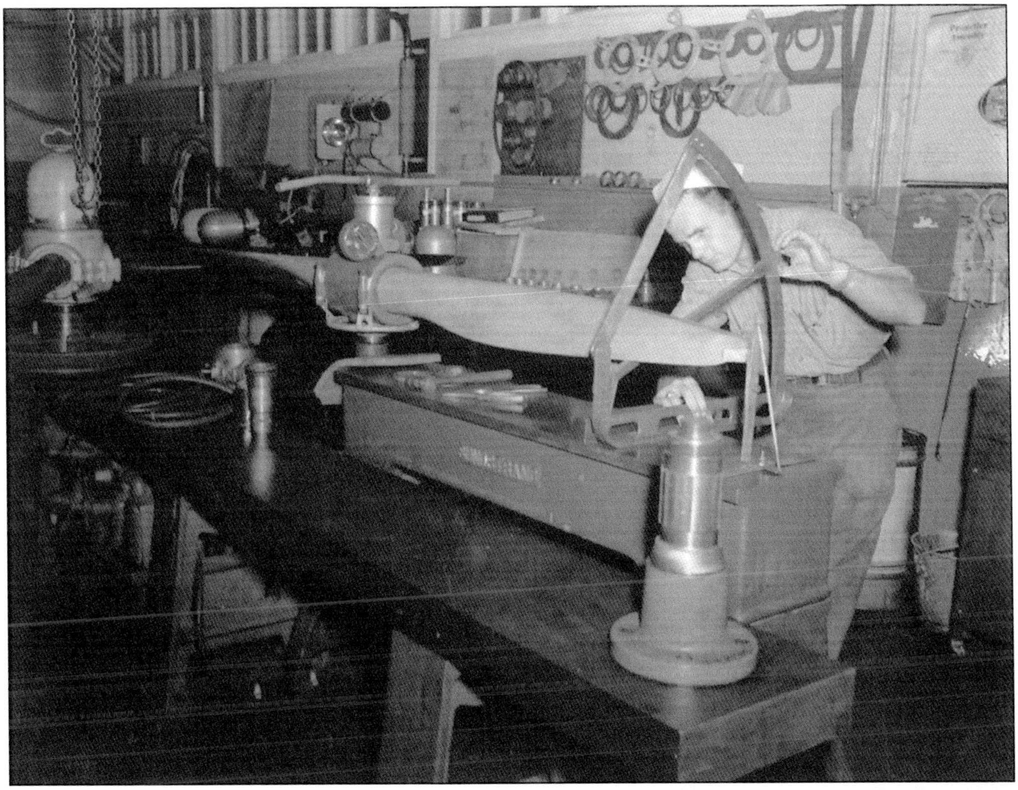

Among the many tasks undertaken at Winter Island was propeller maintenance. By the 1920s, the scientific facts behind the aerodynamics of propellers had been gained. Mechanics knew that nicks and cuts in the wooden blades bore the brunt of the stress created by the spinning motion, and that every effort had to be made to keep the blades as close to perfect as possible. (USCG.)

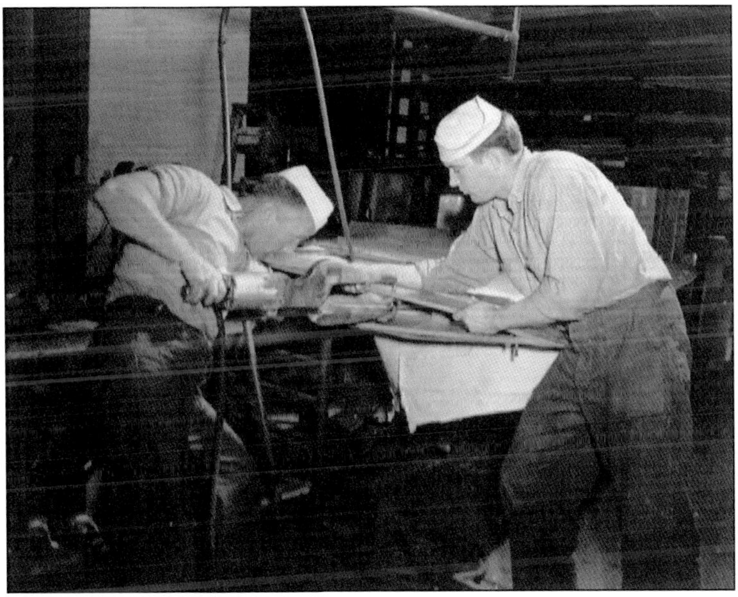

Prior to World War I, during aviation's earliest days, mahogany was the wood of choice for propellers. The war, though, caused shortages of the material and, therefore, experimentation with other woods and other materials. Balancing a prop—making sure that weight was distributed evenly between, in this case, the two blades—was absolutely crucial to safe flight. (USCG.)

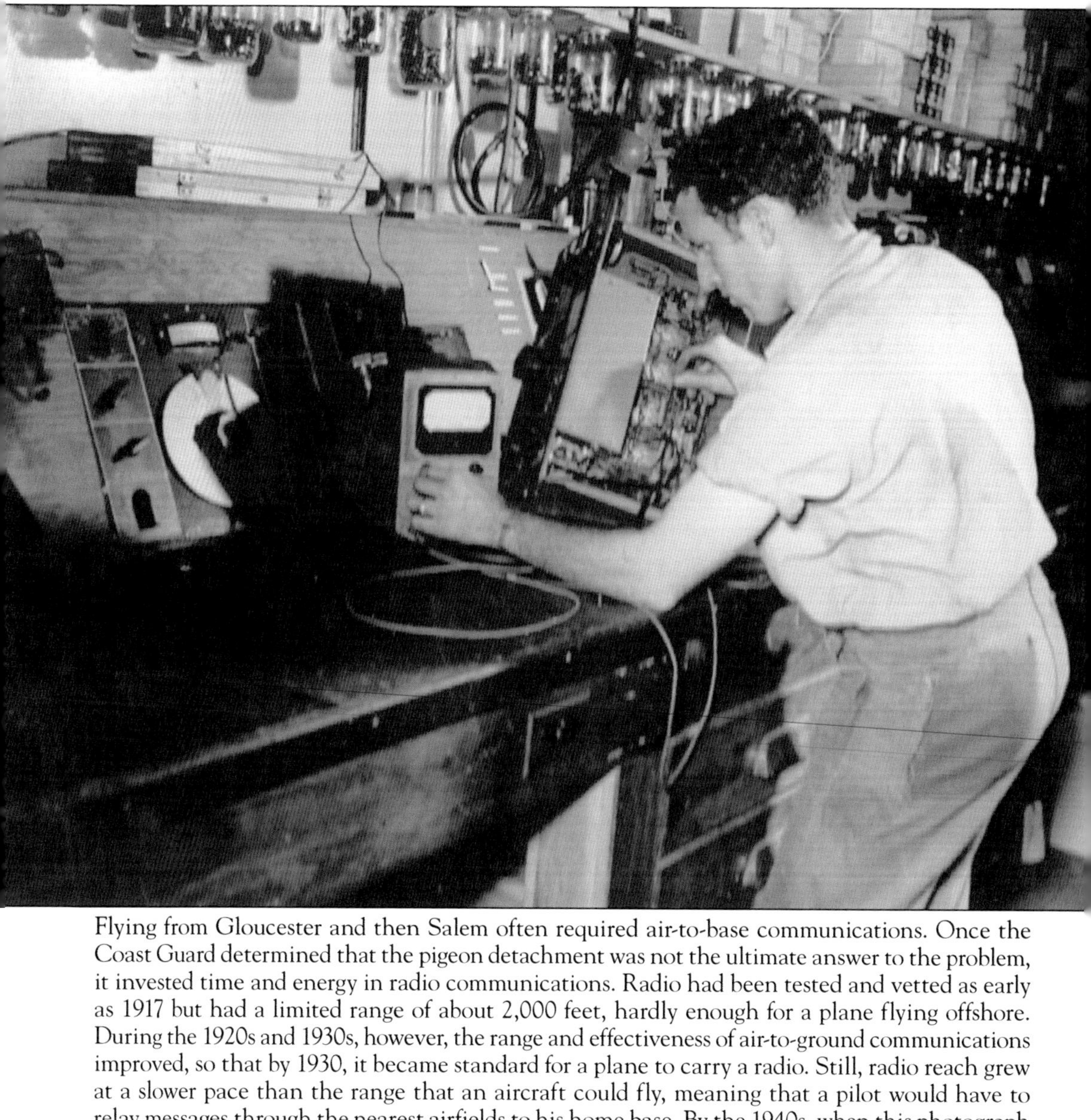

Flying from Gloucester and then Salem often required air-to-base communications. Once the Coast Guard determined that the pigeon detachment was not the ultimate answer to the problem, it invested time and energy in radio communications. Radio had been tested and vetted as early as 1917 but had a limited range of about 2,000 feet, hardly enough for a plane flying offshore. During the 1920s and 1930s, however, the range and effectiveness of air-to-ground communications improved, so that by 1930, it became standard for a plane to carry a radio. Still, radio reach grew at a slower pace than the range that an aircraft could fly, meaning that a pilot would have to relay messages through the nearest airfields to his home base. By the 1940s, when this photograph was taken in Salem, no pilot would go aloft without his radio. It became as important as his parachute. (USCG.)

In the parachute loft at Salem, parachute riggers dried, serviced, and repacked about 110 parachutes every 30 days. Riggers also took care of the station's fabric work of all kinds, which included the repair of life jackets, cargo chutes, foul weather gear, winter gear, and all flight clothing. If it ripped, it hit the parachute loft. (USCG.)

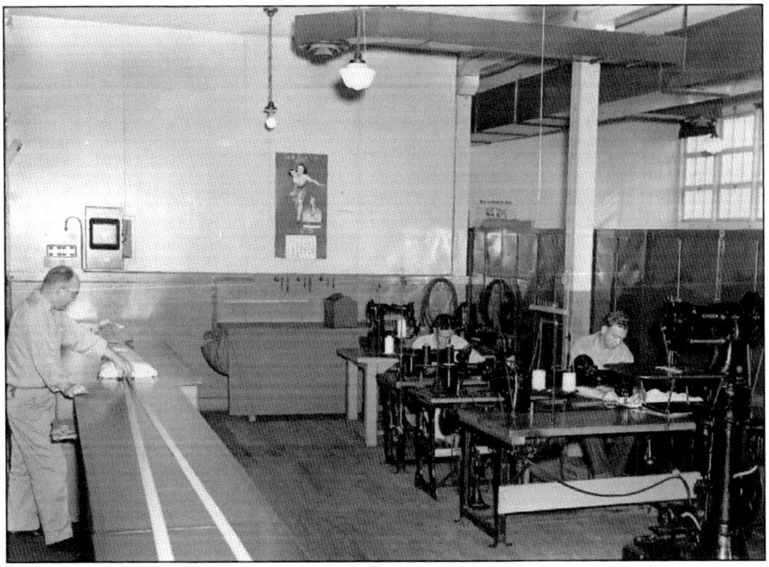

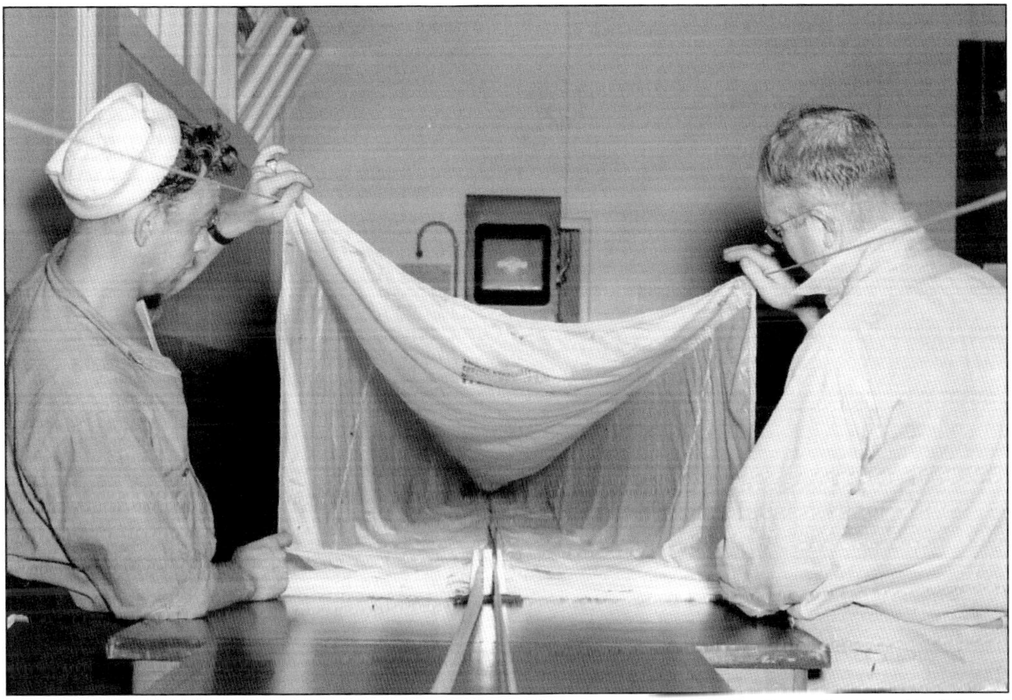

Chief Parachute Rigger LeRoy Spencer, on the right, is joined by Parachute Striker Carl Lund in folding a parachute prior to packing. Parachutes had been in the human mind since the Renaissance, but the first military usage took place in World War I by observers on tethered balloons. After World War I, the US Army developed the modern template of the pilot's chute, including the ripcord and the soft storage pack. (USCG.)

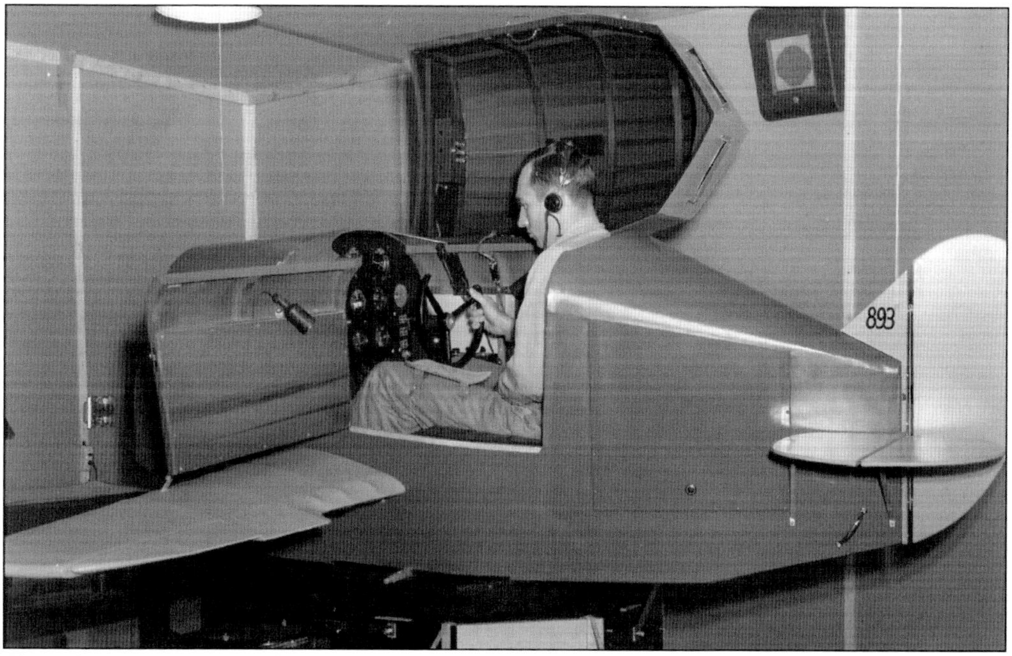

Flying by instruments, without the use of visible navigation aids, was crucial for pilots intending to fly at night or in storms. Here, patrol plane captain Lt. (jg) Arthur Hancock sits at the controls of Air Station Salem's Link Trainer. Ed Link developed the trainer in 1929, and by World War II it was used globally by almost all combatant nations. (USCG.)

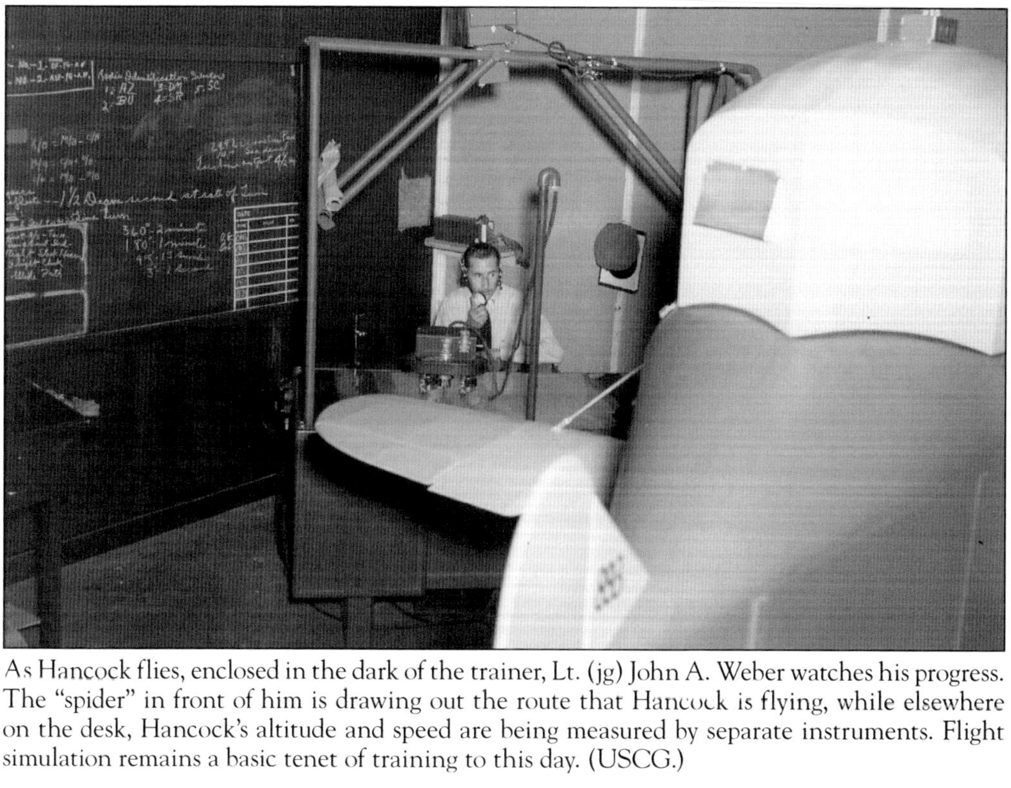

As Hancock flies, enclosed in the dark of the trainer, Lt. (jg) John A. Weber watches his progress. The "spider" in front of him is drawing out the route that Hancock is flying, while elsewhere on the desk, Hancock's altitude and speed are being measured by separate instruments. Flight simulation remains a basic tenet of training to this day. (USCG.)

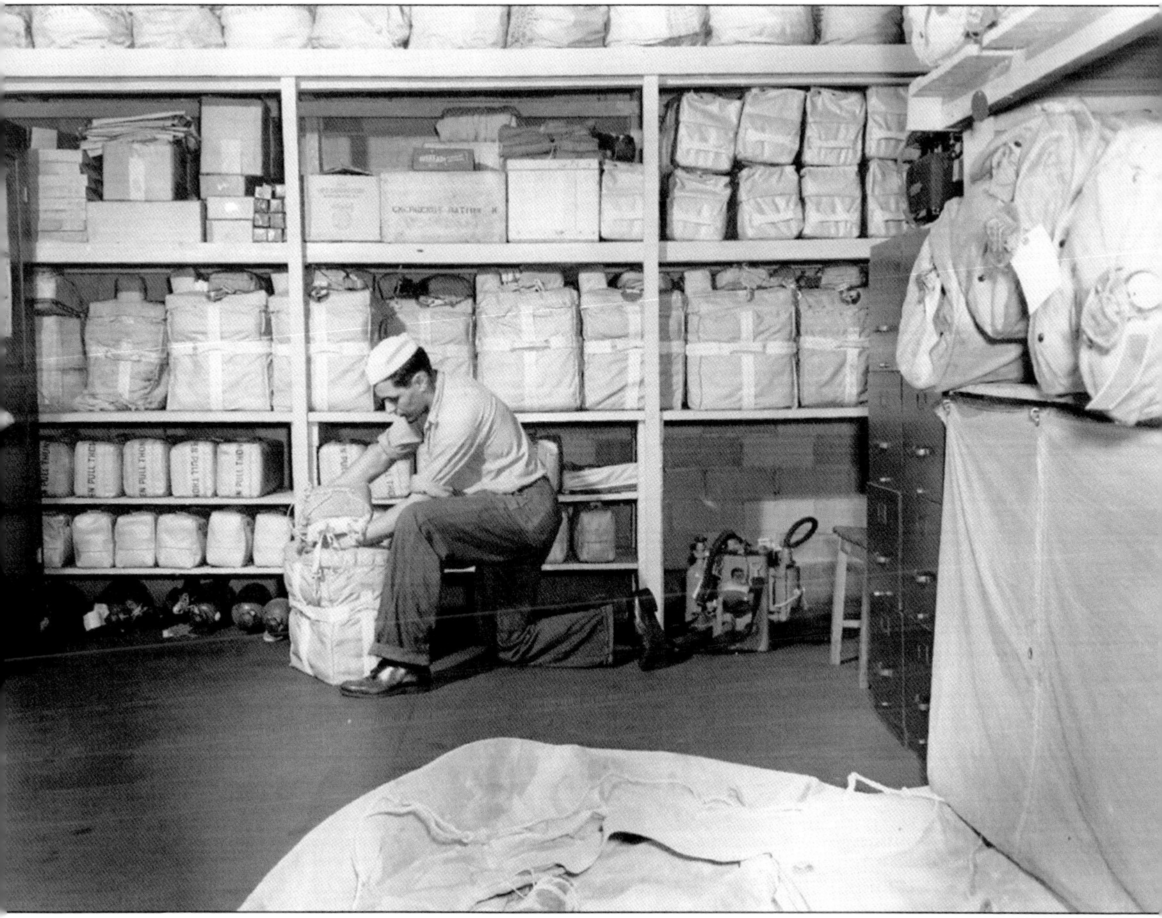

The storage area at Air Station Salem was stocked with everything a pilot wanted but hoped he never needed. Life rafts could be deployed in the case of absolute emergencies. The planes flying out of the station were designed for water landings; should anything catastrophic happen and the plane become uninhabitable or even lost, the life raft would be a last option. In those life rafts would be Gibson Girl kites, bright yellow box kites that came with radios, antennae, and even balloons for windless days. Backing AOM (aviation ordnanceman) 1st Class Albert Neves on the shelves are float flares, oxygen, extra batteries for flashlights, emergency rations, and more. The thought of floating aimlessly on the Atlantic Ocean was not a pleasant one. Pilots who undertook long flights over its vastness knew they would need every possible advantage to be found, from visibility to the ability to survive enough time for search crews to find them. Here, Neves is securing a parachute to a Gibson Girl. (USCG.)

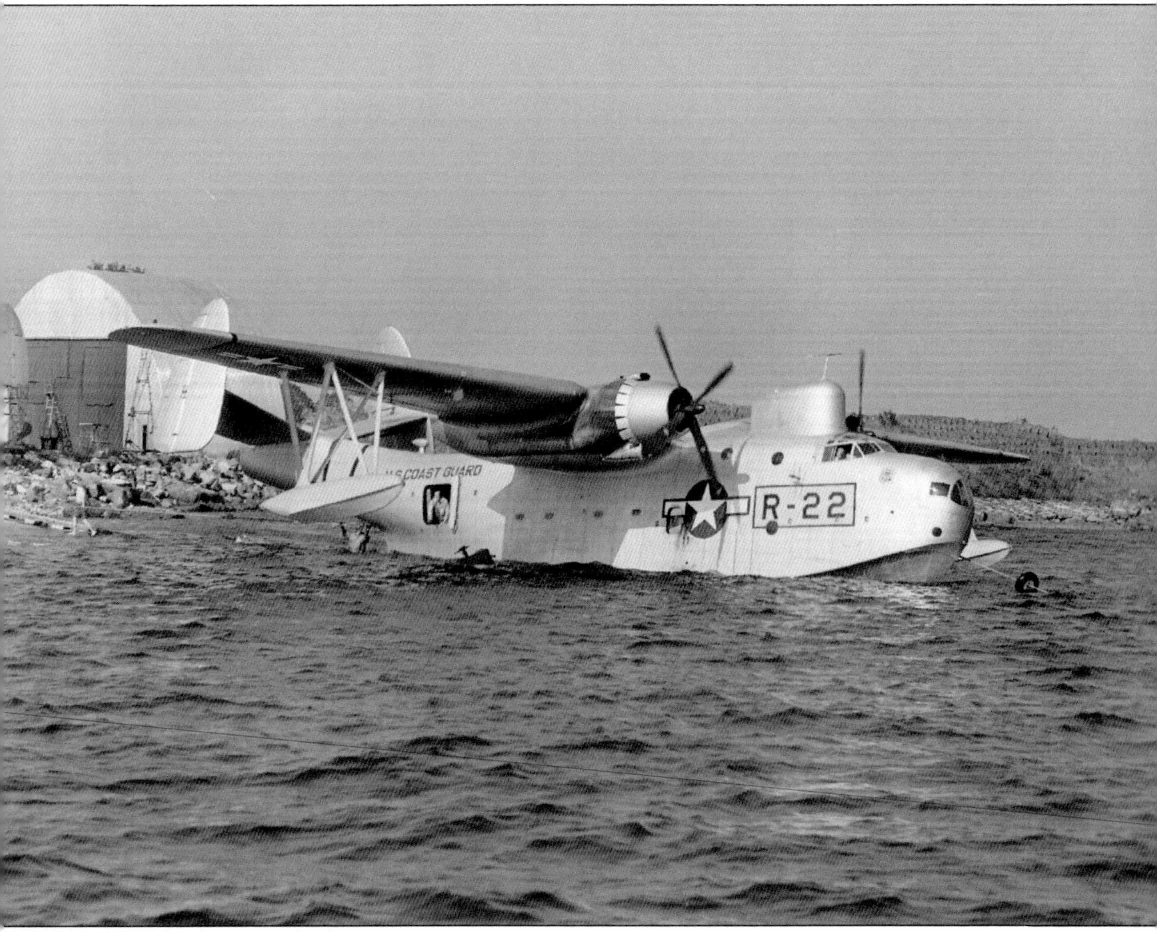

During the first half of 1943, the Coast Guard acquired 27 Martin PBM-3s, and late the next year the service acquired 41 PBM-5 models. More came in the latter half of 1945. For the next decade, they would serve a major role as the long-range aerial search and rescue aircraft of the service. By 1956, they were gone from service. According to the Coast Guard, "As of 31 October 1944 the Coast Guard's PBM's were stationed at the following Coast Guard Air Stations: five PBM-3's at Elizabeth City; one PBM-3C at Miami; two PBM-3S's at Port Angeles; two PBM-3S's at St. Petersburg; two PBM-3S's at Salem; one PBM-3C and three PBM-3S's at San Diego; and two PBM-3S's at San Francisco." Kenneth E. Euritt, who flew on Martins out of Salem from 1952 to 1955, said in 2008, "I can close my eyes and still hear the roar of the engines and the sight of the plane being engulfed in spray as we came up on the step in order to get to flying speed." (USCG.)

Winter Island is not an island, in that it is connected to the mainland of Salem by a causeway. When approaching the gate to the Coast Guard station, motorists were required to slow down to 10 miles per hour and were instructed that there would be "No Driving on Aprons." Winter Island had a long history of military usage, dating back to the first iteration of Fort Pickering in the 17th century. (USCG.)

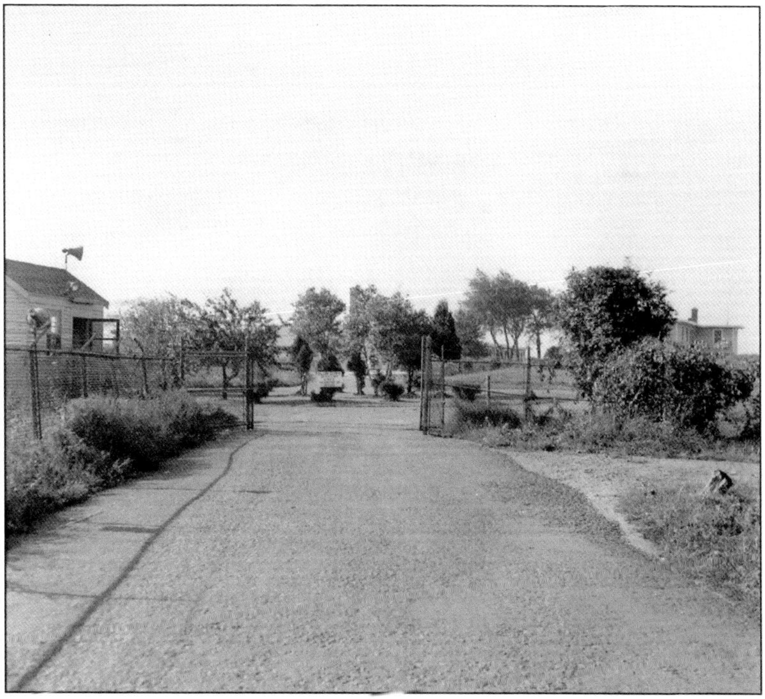

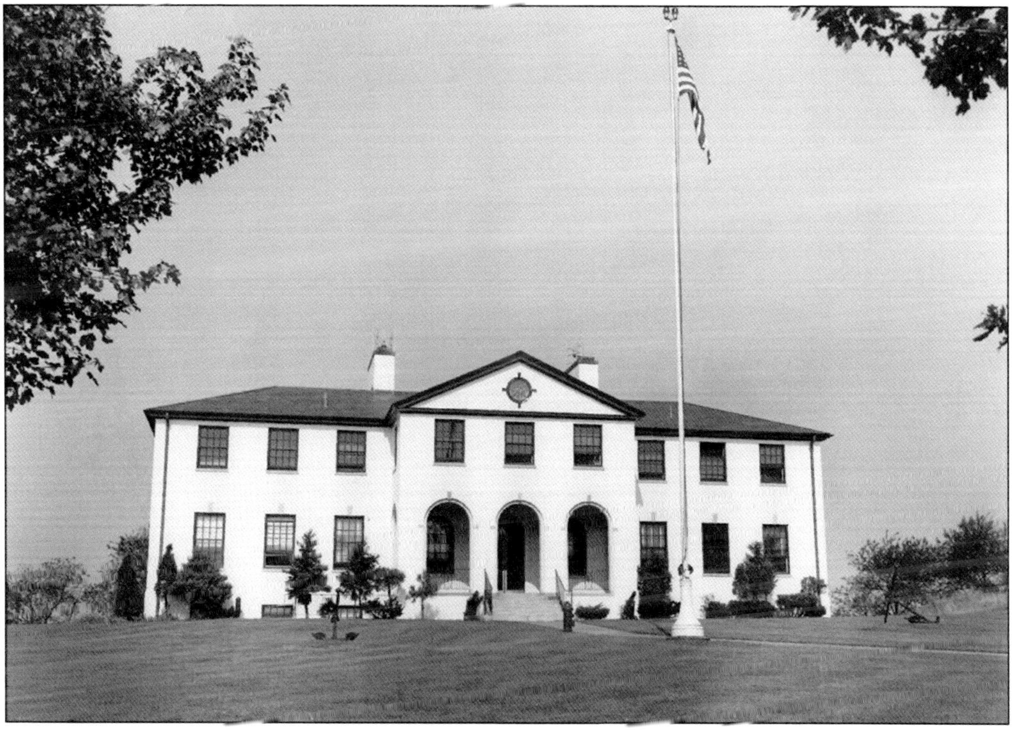

The Coast Guard's impressive barracks at Winter Island had all the accoutrements: a cannon, anchors resting on the lawn, and a majestically tall and patriotic flagpole. The building today lies nearly in ruins after 46 years of disuse. In its halls echo the voices of Coast Guardsmen talking about rescues, schools of fish sighted, the news of the war, Red Sox scores, and more. (USCG.)

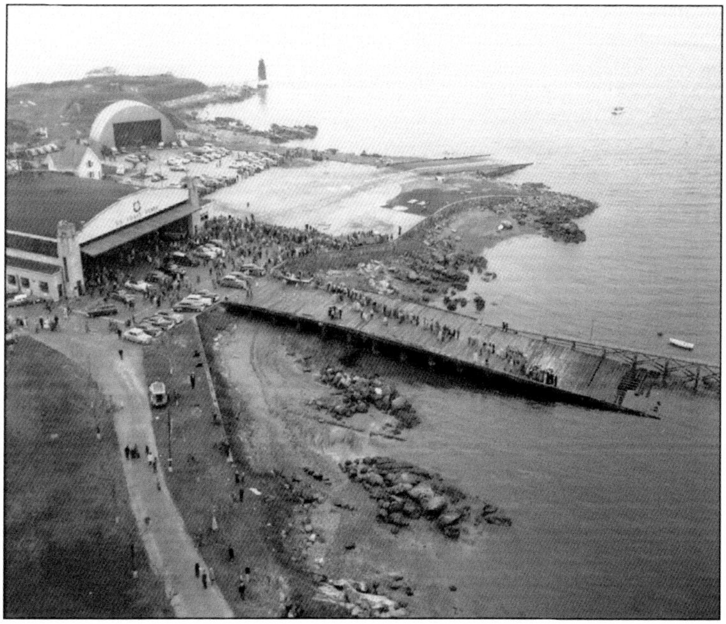

Occasionally, to celebrate an anniversary like Coast Guard Day (August 4), the service staged an open house for the public. During those occasions, the local citizenry witnessed demonstrations of the Coast Guard's rescuing prowess. Airmen showed off the latest planes and their capabilities, describing how they are best utilized. For the fishing communities north of Boston, the knowledge was more than simply entertaining. (USCG.)

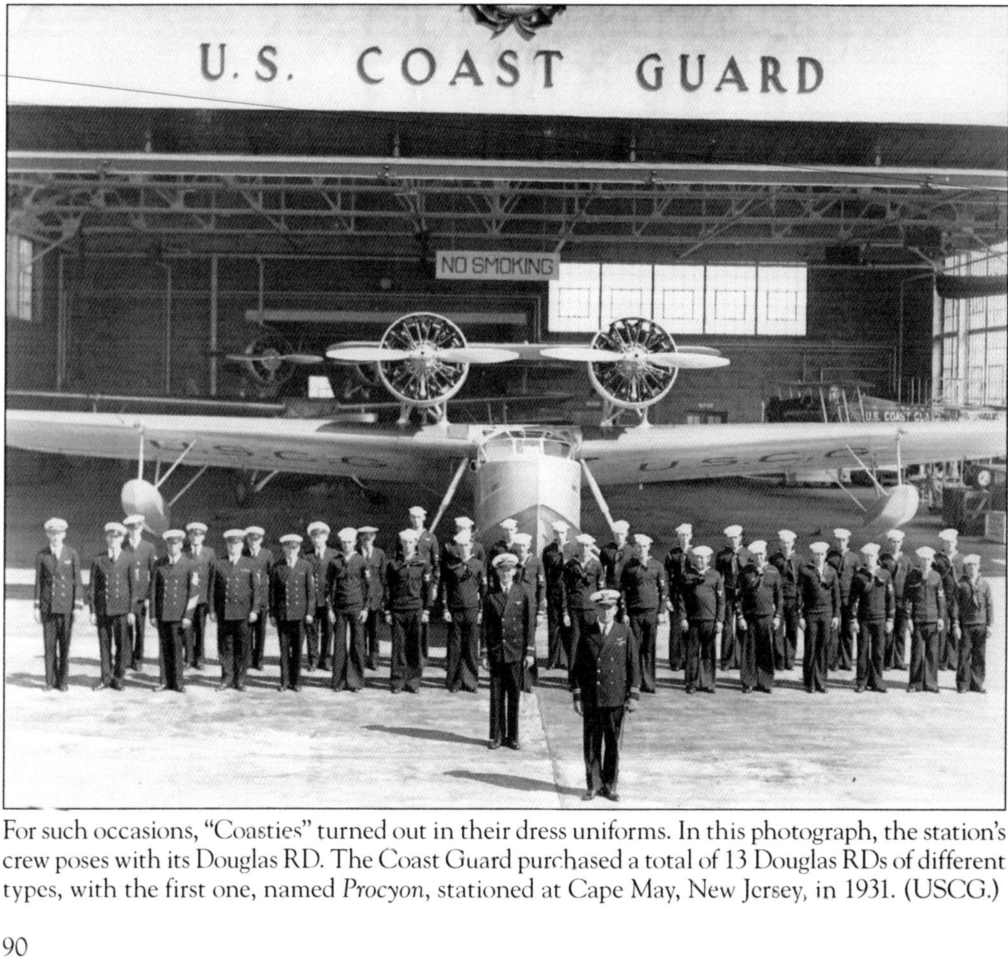

For such occasions, "Coasties" turned out in their dress uniforms. In this photograph, the station's crew poses with its Douglas RD. The Coast Guard purchased a total of 13 Douglas RDs of different types, with the first one, named *Procyon*, stationed at Cape May, New Jersey, in 1931. (USCG.)

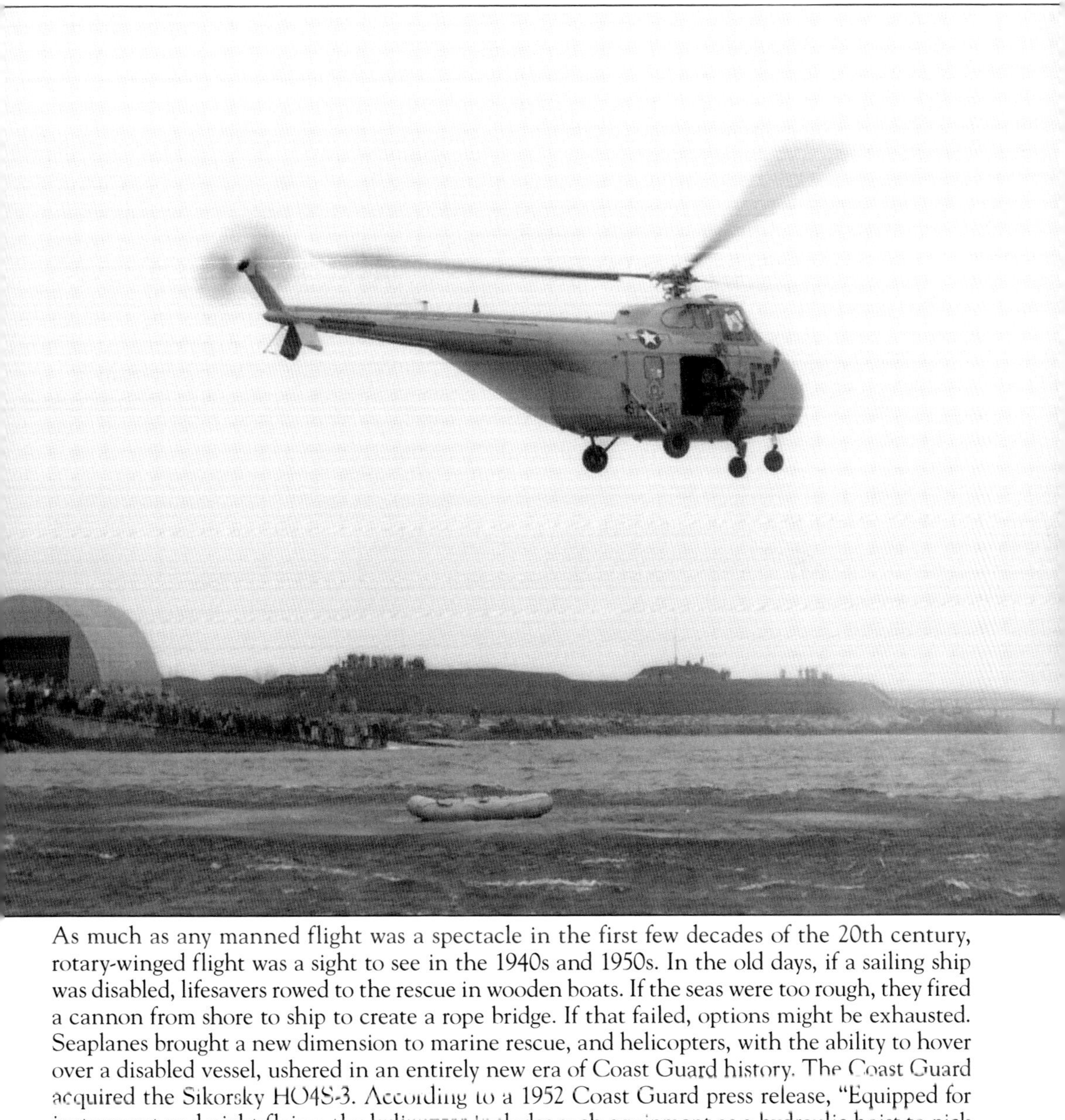

As much as any manned flight was a spectacle in the first few decades of the 20th century, rotary-winged flight was a sight to see in the 1940s and 1950s. In the old days, if a sailing ship was disabled, lifesavers rowed to the rescue in wooden boats. If the seas were too rough, they fired a cannon from shore to ship to create a rope bridge. If that failed, options might be exhausted. Seaplanes brought a new dimension to marine rescue, and helicopters, with the ability to hover over a disabled vessel, ushered in an entirely new era of Coast Guard history. The Coast Guard acquired the Sikorsky HO4S-3. According to a 1952 Coast Guard press release, "Equipped for instrument and night flying, the helicopter includes such equipment as a hydraulic hoist to pick up personnel or equipment up to 400 pounds in weight. . . . Plans also have been drawn up for the installation of a collapsible rubber float attached to the landing gear. . . . The new helicopters are painted bright yellow, and are the largest helicopters operating in the Coast Guard." (USCG.)

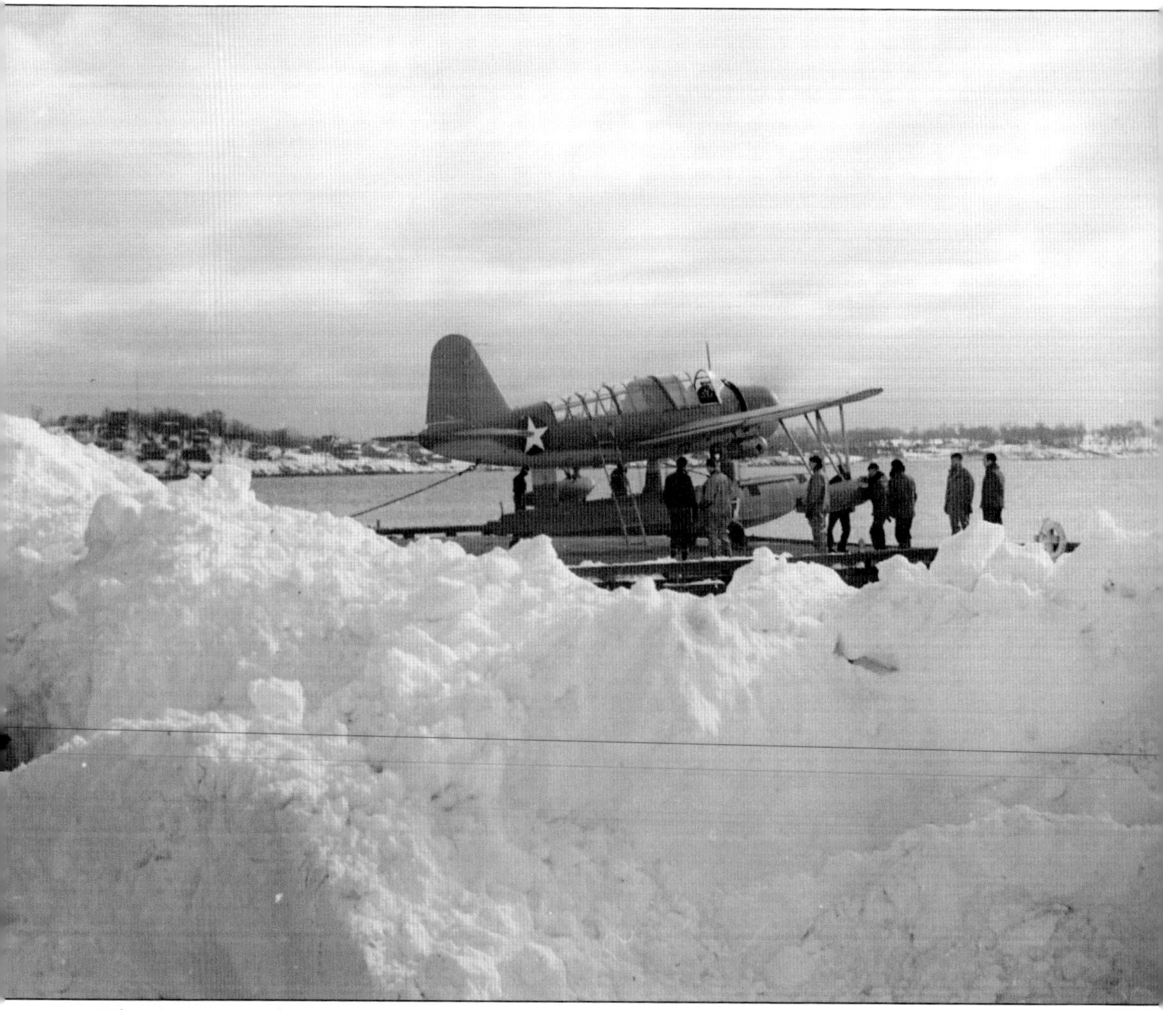

The Coast Guard's roots run deep in Massachusetts. Formed by several predecessor agencies, the service can trace its founding to several places. The first American lighthouse was constructed on Little Brewster Island, Boston Light. The first revenue cutter was launched in Newburyport. The nation's system of lifeboat stations was based on the work of the Humane Society of the Commonwealth of Massachusetts in the 1800s. Coast Guardsmen operating out of Air Station Salem knew that they might be called upon to put their lives in danger to protect others from death upon the seas. They knew that they might have to brave extreme cold, high winds, and even waves clawing upward at their aircraft. But, like these men with their OS2U-3 Kingfisher, they all knew that this was the job for which they had signed up. Today's Coast Guard motto, *Semper Paratus*, or "Always Prepared," was as evident then as it is in the 21st century. (USCG.)

For 35 years, Salem was the Coast Guard's Massachusetts home. Crews flew from the base up to Newfoundland, far out to sea, and even westward to Pennsylvania when the call for help rang out. Knowing that at times local weather conditions could hinder or even halt flight operations, the command opened a secondary station for added support. (USCG.)

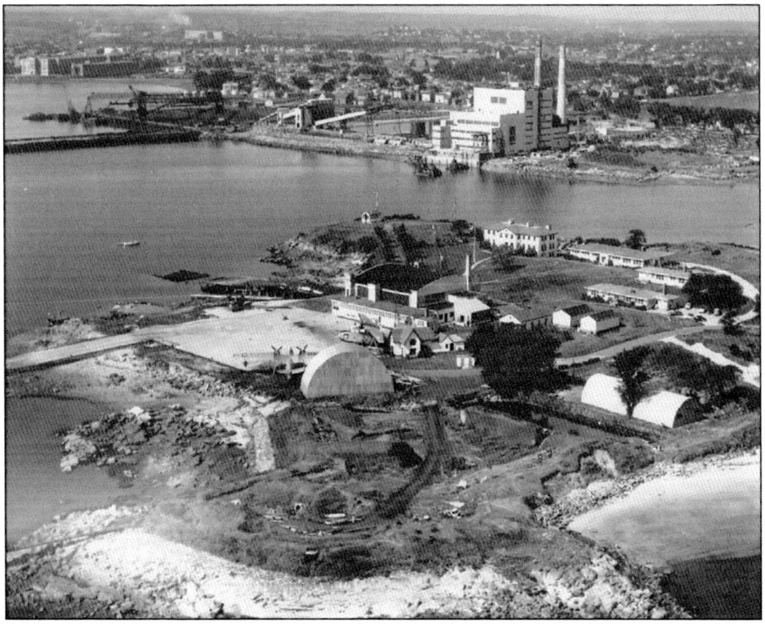

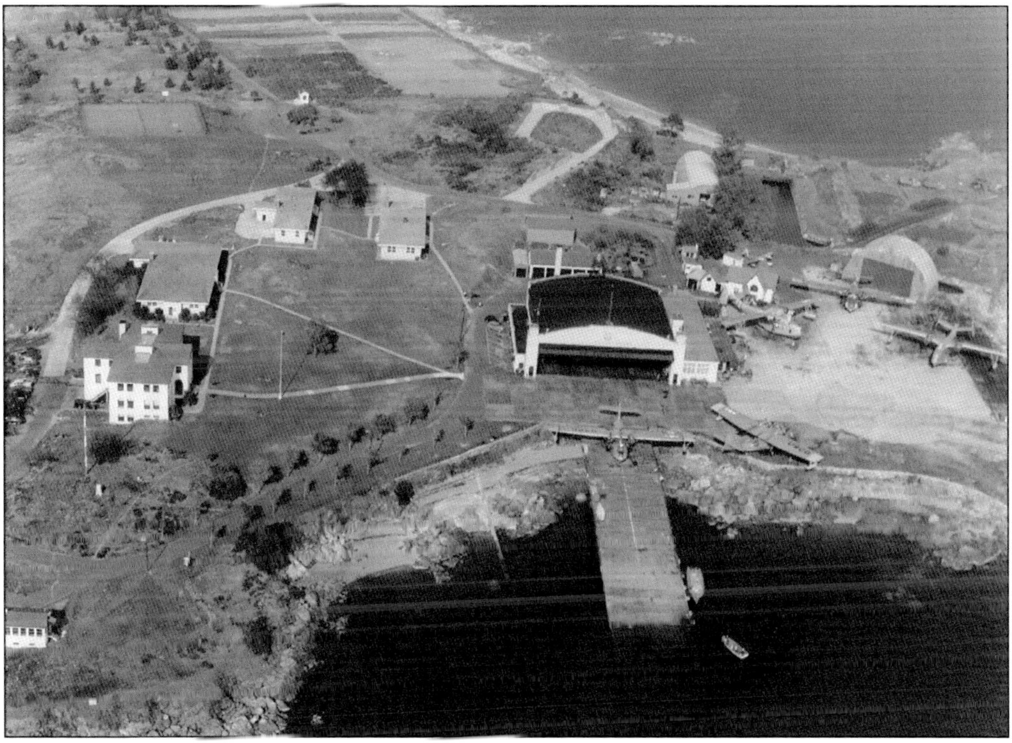

That station, Coast Guard Air Detachment Quonset Point, boasted just a single plane, four pilots, and eight crewmen. The crew there worked in what are known as port and starboard duty sections, two groups working on a schedule of three days on and three days off. The main force, though, remained here, at Air Station Salem. (USCG.)

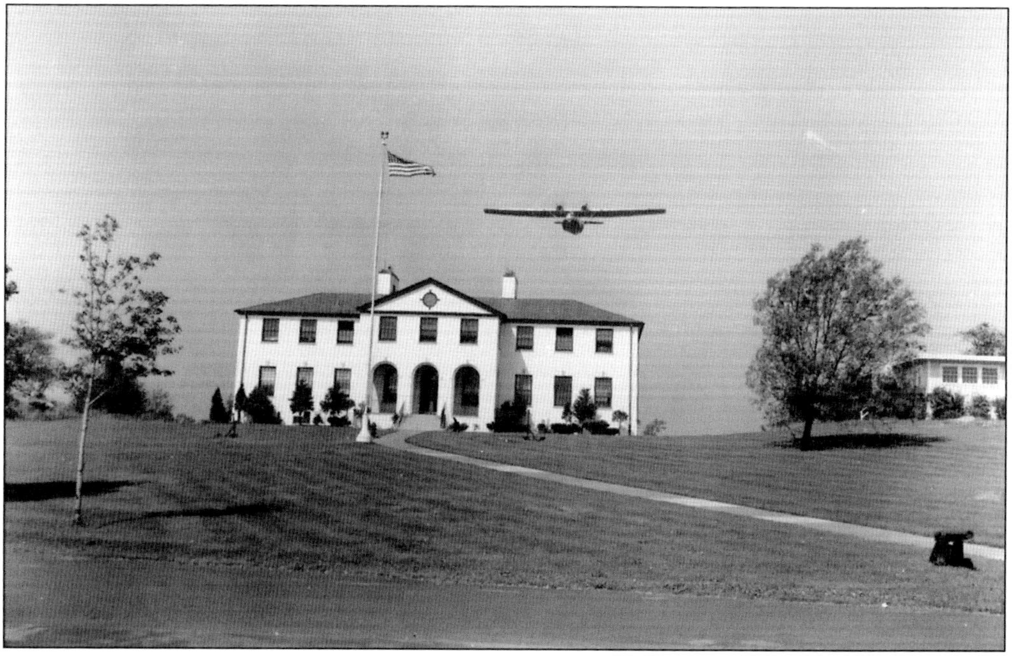

By the late 1960s, Air Station Salem was nearing the end of its operational life. Scenes like this one would soon fade into memory. The station closed in 1970, taking its substation at Quonset Point with it. Both units would be consolidated into a new command on Cape Cod, taking over the old Otis Airfield on what is now Joint Base Cape Cod. (USCG.)

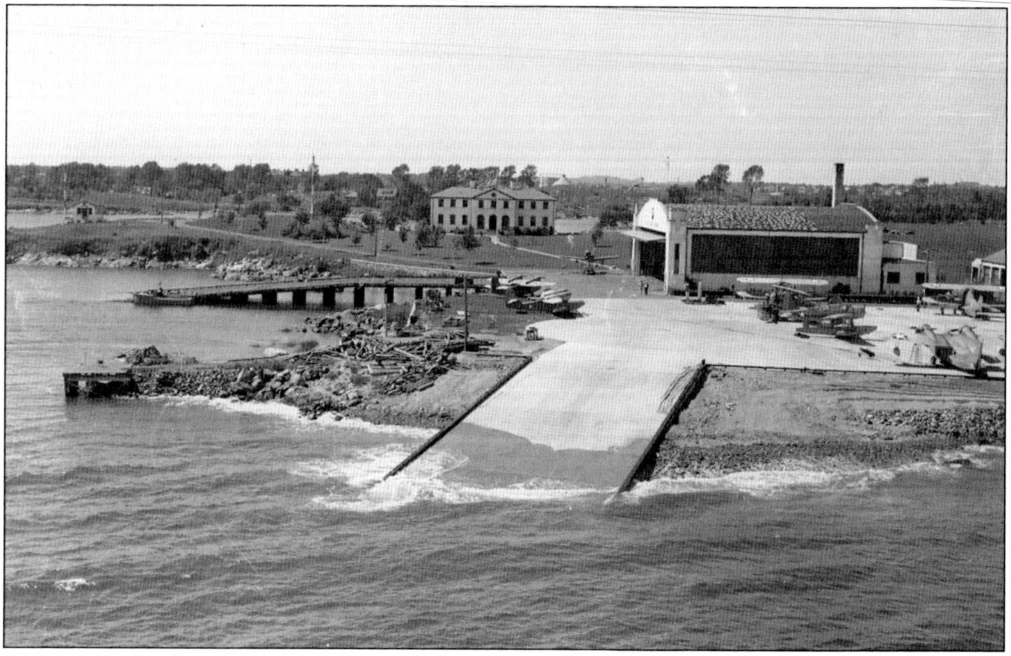

Since 1970, the hangar and the other buildings at Winter Island have fallen into disrepair. The Coast Guard turned the property over to the City of Salem in 1972. The island now hosts Winter Island Marine Park and is listed in the National Register of Historic Places. In 2011, Salem developed a master plan for the future of the island. (USCG.)

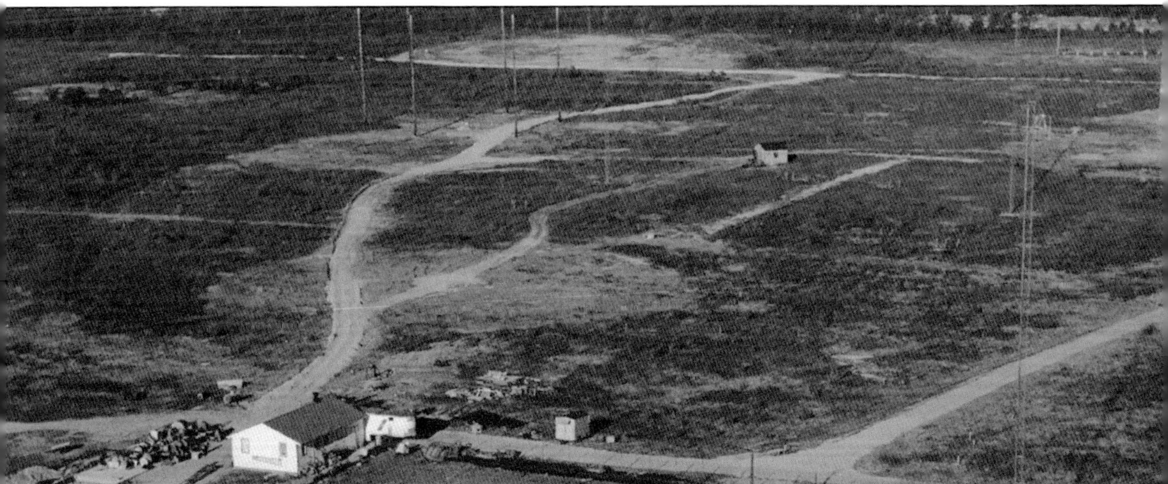

The slow move away from floatplanes meant that the Coast Guard, ironically, could move away from the coast. As helicopters became more prominent, with the ability to lift off vertically from just about anywhere, shoreside air stations became obsolete. And so, in 1970, the Coast Guard consolidated its Salem and Quonset Point operations in the sandy pine barrens of Cape Cod. The Otis Air Force Base started as the landing field of Camp Edwards, built during the Depression as the training ground for the state's military forces, replacing Fort Devens in Central Massachusetts. Planes and crews from Otis played major roles in the Cold War. In 1968, the Department of Defense okayed the opening of Air Station Cape Cod. The HH-3F Pelicans and HU-16E Albatrosses then at Salem and Quonset Point moved to Cape Cod for an August 29, 1970, commissioning of the new station. (USCG.)

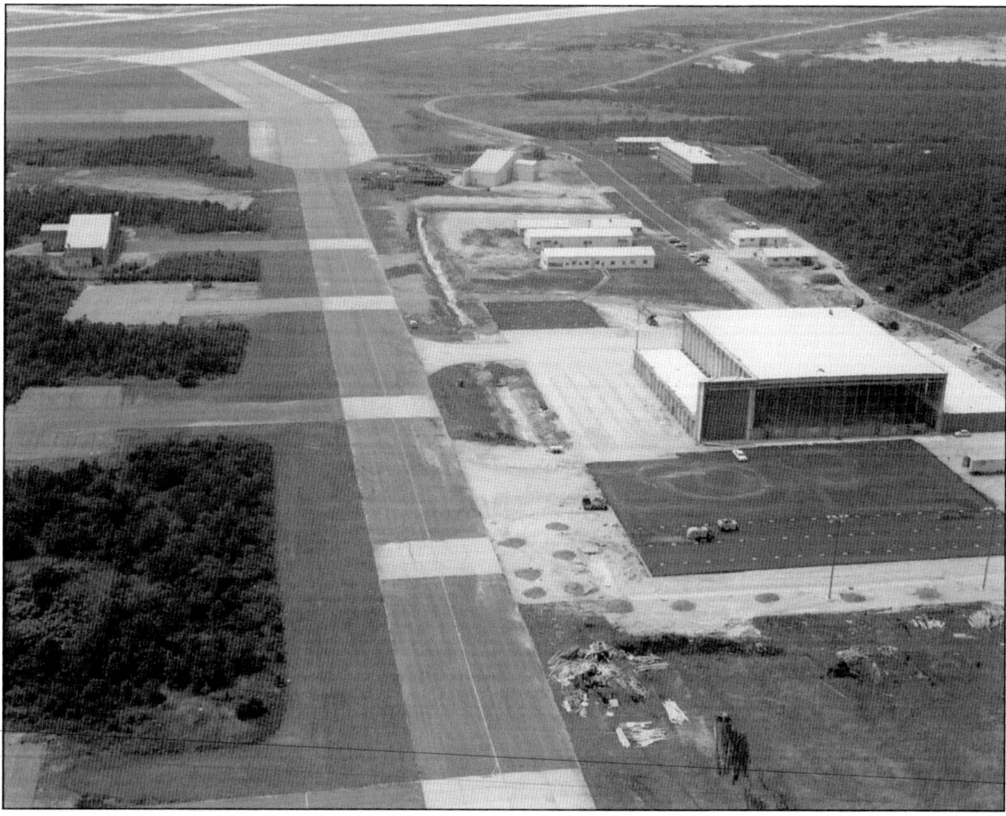

The move to the new base opened avenues to search and rescue for the Coast Guard, as it could now explore jet operations in the area. The Coast Guard, though, held tightly to its amphibians. HU-25 Guardians reached the base in 1982, which sounded the final bell for the much-loved Albatrosses. The last one, CG-7250, was the last fixed-wing amphibious aircraft in the US inventory when it was retired on March 10, 1983. Today, it is on display outside the entrance to the base. (USCG.)

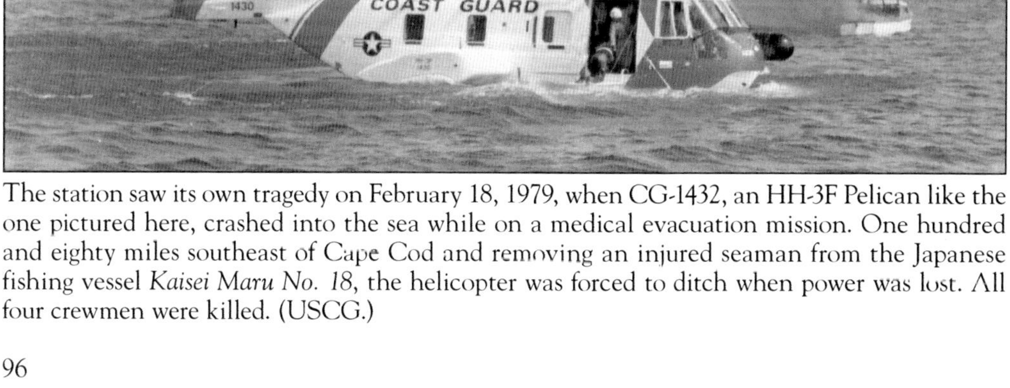

The station saw its own tragedy on February 18, 1979, when CG-1432, an HH-3F Pelican like the one pictured here, crashed into the sea while on a medical evacuation mission. One hundred and eighty miles southeast of Cape Cod and removing an injured seaman from the Japanese fishing vessel *Kaisei Maru No. 18*, the helicopter was forced to ditch when power was lost. All four crewmen were killed. (USCG.)

Five

INNOVATORS AND EVENTS

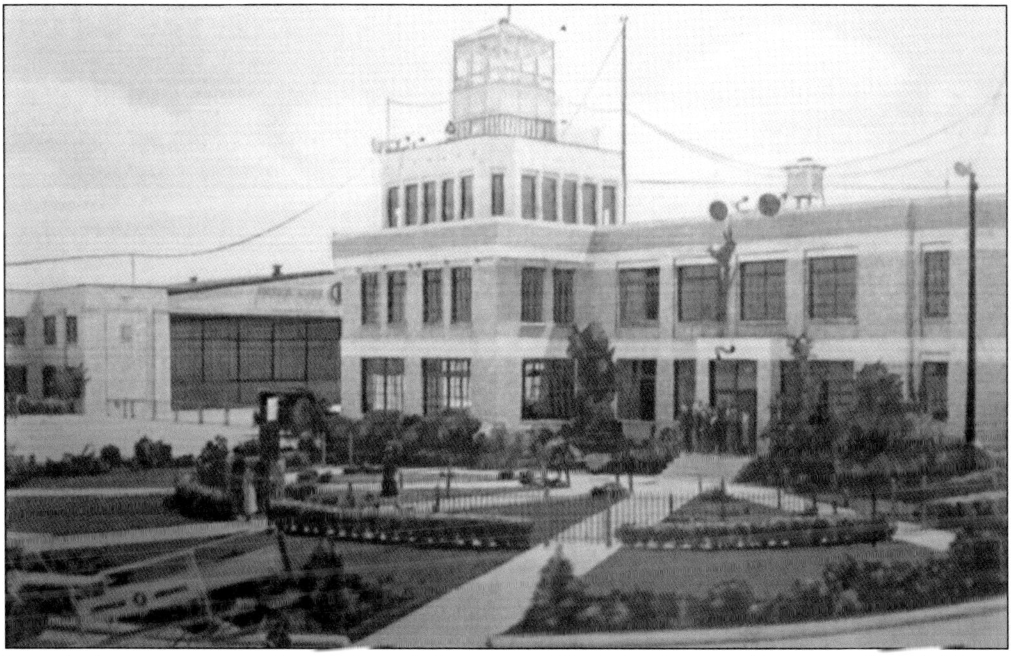

Long before international arrivals and departures filled the Boston skies, a small airfield was begun in East Boston. By the mid-1920s, hangars representing Curtiss Aircraft, Skyways, and East Coast Aircraft began dotting the field. By the 1950s and 1960s, mud flats and tidal lands were being filled in to expand the East Boston airport into the current Logan International Airport, capable of handling large-capacity passenger and cargo jet aircraft. (MAHS.)

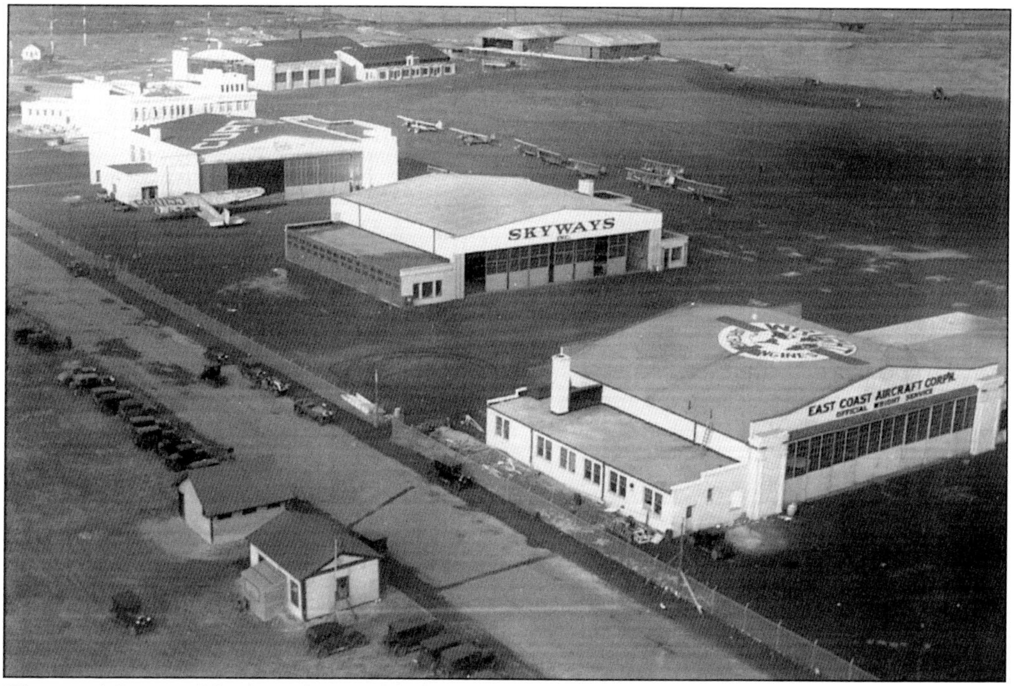

The word "runway" had not been added to aviation terminology when the East Boston airport opened as Jeffrey Field on September 8, 1923. In the early years, paved, directional runways were not yet employed, and the takeoff was made from a "get-a-way" by pointing the plane into the wind for maximum lift and then going for it. (MAHS.)

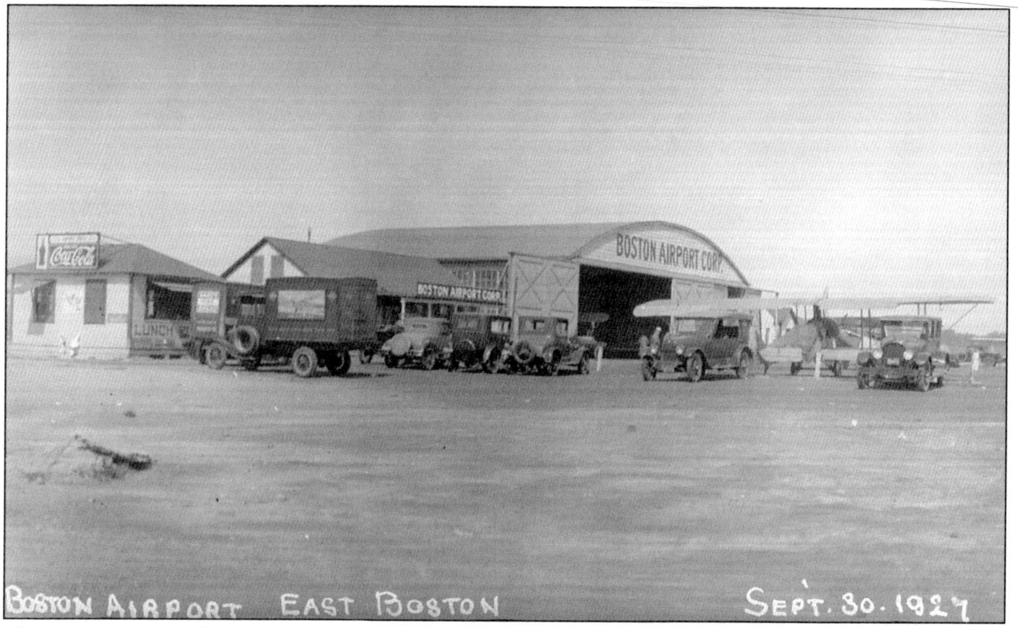

The first commercial airline passengers to travel from Boston to New York left in 1927 aboard Colonial Air Transport planes. Juan Trippe, the entrepreneur who invested in the line and moved it to Boston, would also found Pan Am and appear on the cover of *Time* magazine in 1933. In 1936, the airport became the official weather tracking station for the National Weather Service. (MAHS.)

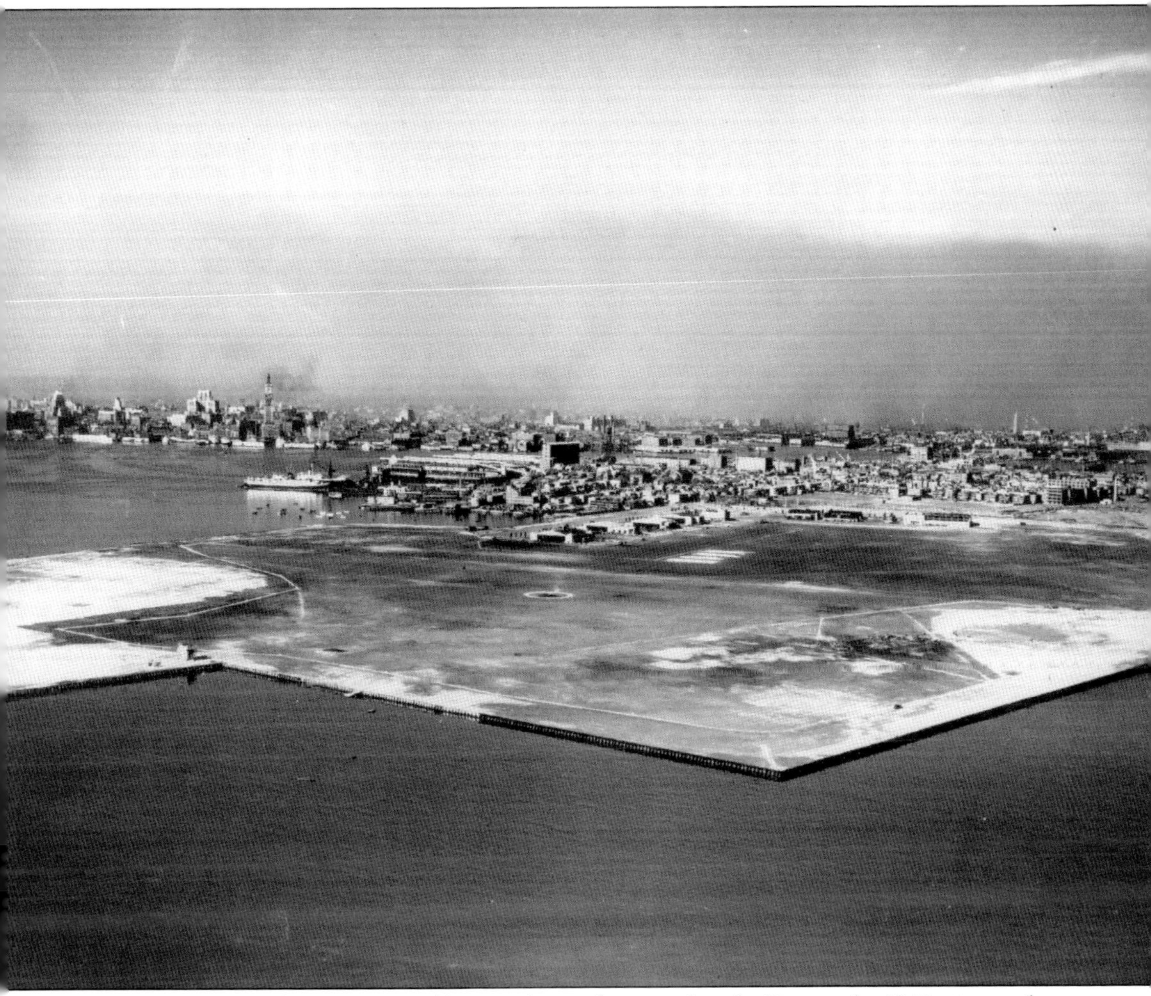

Shown here in 1938, the airport would expand over the next decade. During the 1940s, it grew by 1,800 acres as some of the smaller Boston Harbor islands—Bird, Governor's, Noddle's, and Apple—were cut down to create landfill for the expansion, at that point fading into history. (MAHS.)

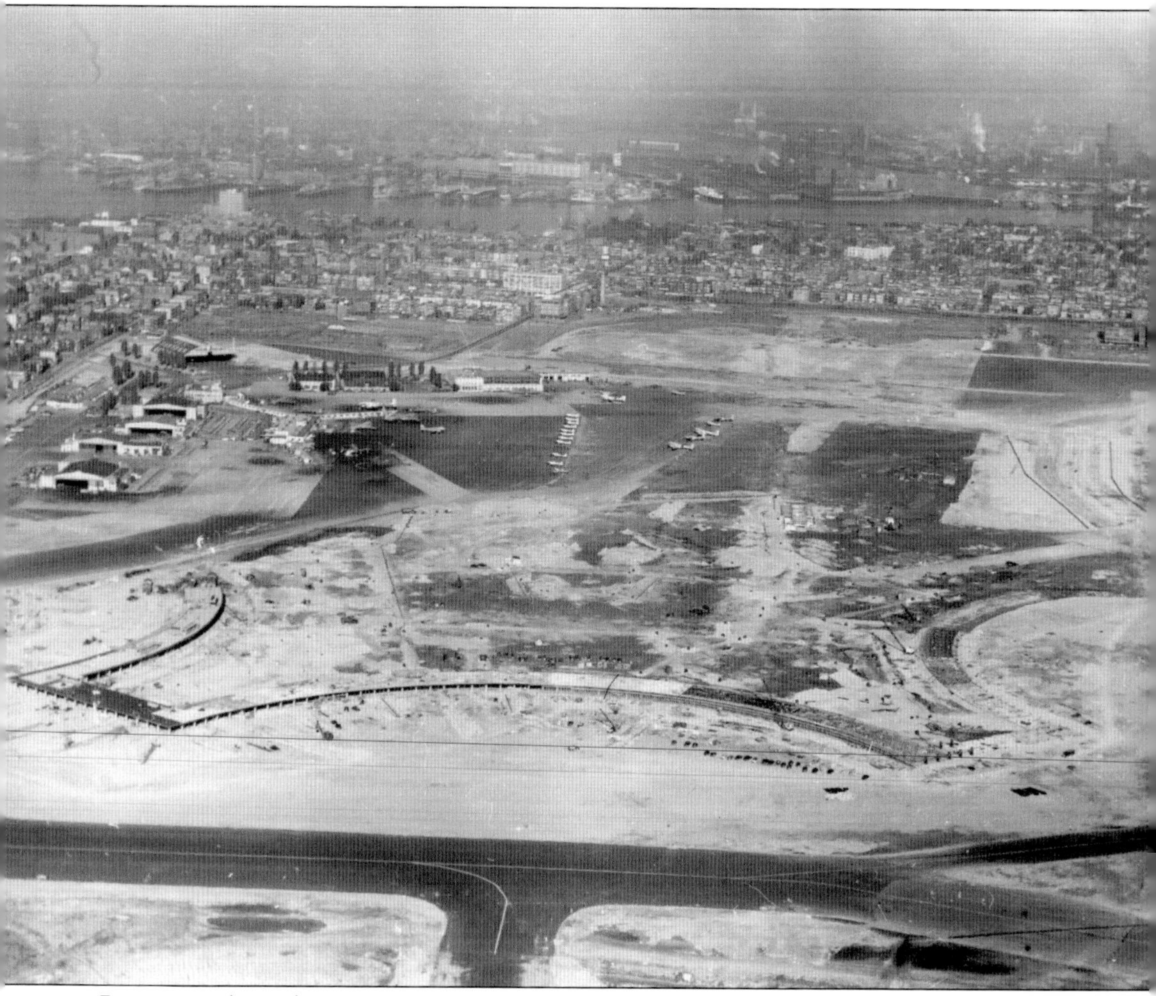

By 1948, as shown here, Logan International Airport was a grander place than it had been when the first planes rumbled across the fields and into the sky. In 1944, just two airlines operated from the airport: Northeast flew to the northern New England states and Canada, and American Airlines handled traffic to New York. By 1949, the airport was hosting 471,000 passengers annually. In 1952, Logan became the first airport in America linking air to rail with the opening of the Airport station on the MTA's Blue Line, and in 1956, the Massachusetts Legislature created the Massachusetts Port Authority (Massport) to oversee the operations of Logan. By 1959, Logan had four runways and 45 gates at an expanded terminal. Jets began flying to Europe daily that year and then to Los Angeles. The airport became more than just a local airfield, it became New England's regional airport, the hub from which the residents of the six-state region flew when they wanted to explore the world. (MAHS.)

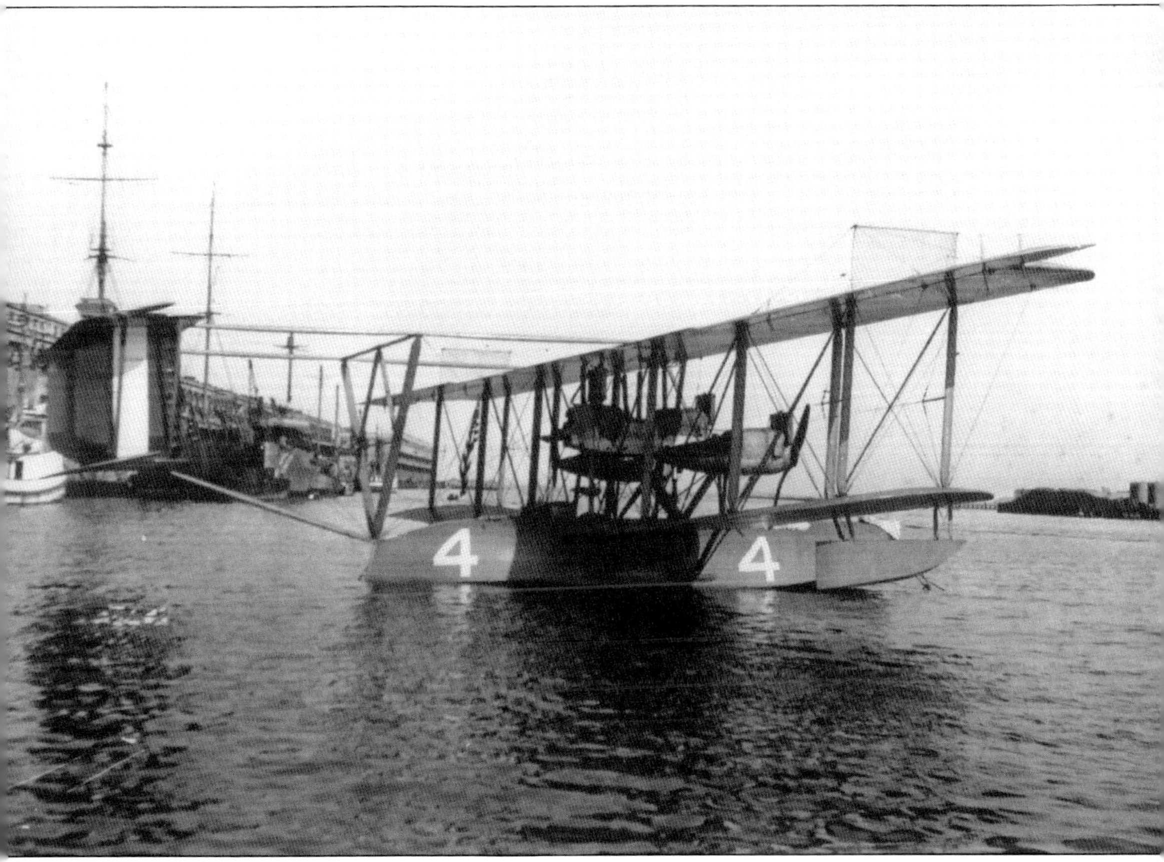

Albert Cushing "Putty" Read was born in New Hampshire into a Boston Brahmin family that eventually moved to Hanson, Massachusetts. His stoic and steadfast ways earned him his nickname during his Navy years. Unflustered by any situation, Read rarely showed any emotion. His face, his comrades said, was like putty. With such determination and evenness of temperament, Putty became the perfect choice for a daring aviation mission. Entering World War I in 1917, the United States ordered built its first four-engine planes, Navy-Curtiss (NC) flying boats, with the notion of flying them across the Atlantic for combat. But the war ended before they could take to the skies. Then, at peace, the Navy decided to make the trek anyway. It would be the first transatlantic crossing accomplished by anything but a boat. Piloting the NC-4 Curtiss Flying Boat to six stops along the way, Lt. Comdr. "Putty" Read accomplished the deed from May 8 through May 31. (MAHS.)

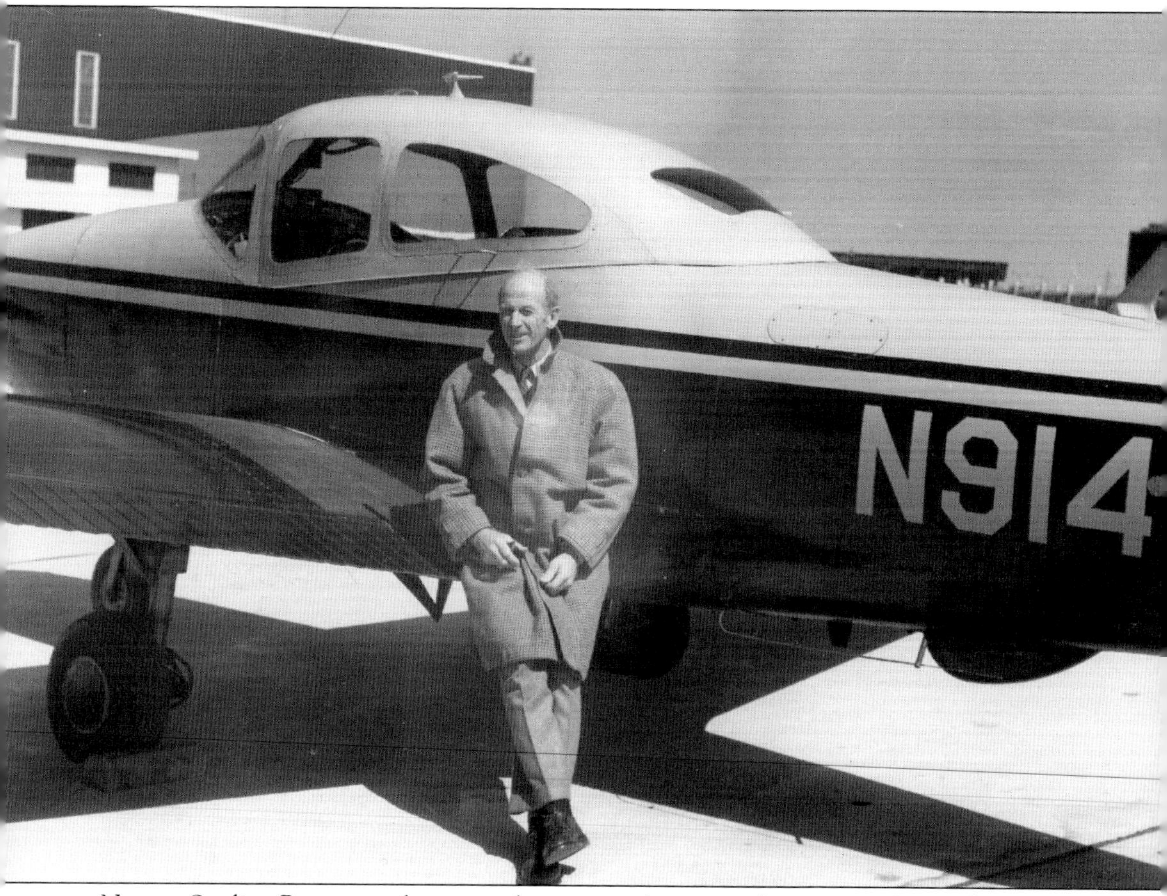

Next to Starling Burgess, perhaps no other person championed aviation in Massachusetts more than Crocker Snow. Born in 1905, Crocker was captivated by the stories he heard of World War I aces and his brother, Bill, a naval aviator who flew antisubmarine patrols off the French and British coasts. Crocker received Massachusetts pilot license no. 5 (signed by Orville Wright) in 1927. He earned money giving rides in an OX-5 Travel Air biplane. One of his first passengers was a social worker at the Denison House in Boston, Amelia Earhart. In 1928, he founded Skyways, one of the first commercial airlines in the state. He served as the first director of aeronautics for Massachusetts in 1939 and, in 1941, took control of the effort to ferry US aircraft across the North Atlantic to Europe. During the war, Crocker commanded the 498th Bombing group in the Pacific. Later, in the 1970s, he was appointed chairman of the Federal Aviation Advisory Commission, created by Congress to study the long-range requirements of commercial and private aviation. (MAHS.)

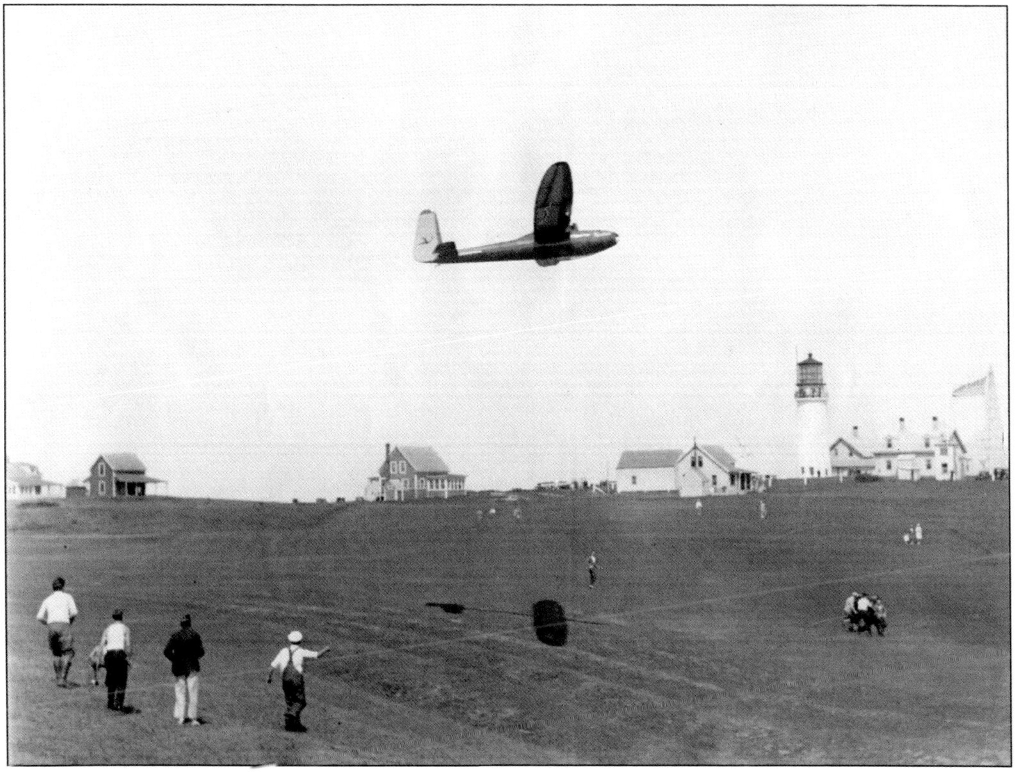

Truro's Highland Light was the site of a record-breaking glider flight on July 29, 1928. Three German pilots traveled to the United States to set up a glider camp, and one of them, Peter Hesselbach, set a new record of time aloft with four hours and five minutes, to the delight of the gathered crowd. When the gliders slid out over the sea, the daring nature of the maneuver sent chills through the observers. (MAHS.)

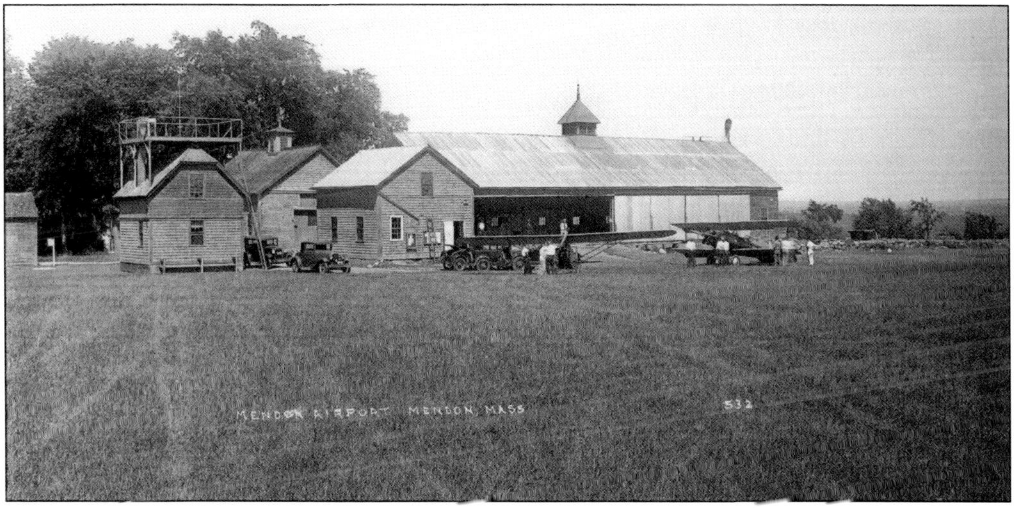

In the 1930s, any city without an airport was considered behind the times. As such, even small towns dreamed big. Mendon, southwest of Boston, had an airfield that today is nearly lost to history. P.A. Millis purchased the property in 1927, but the first official mention of the field is in the 1929 Commercial Department Airway Bulletin. By 1935, it was a commercial field. (MAHS.)

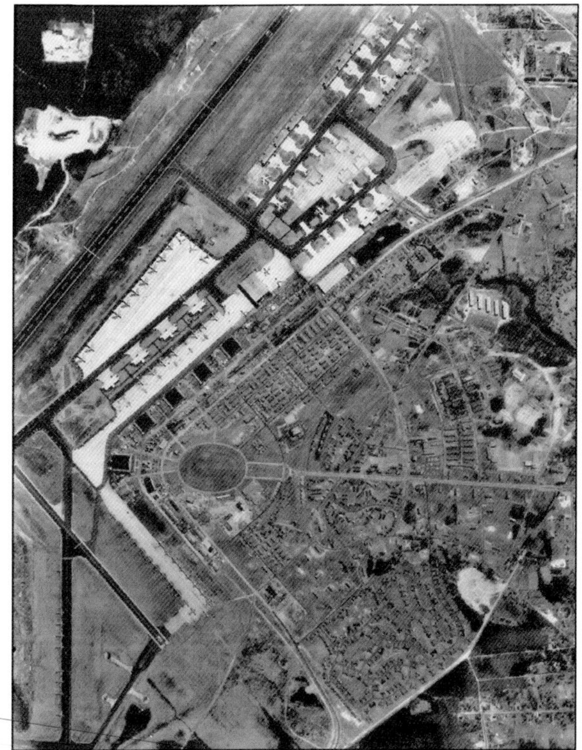

Westover Air Reserve Base opened in Chicopee and Ludlow in 1940 in anticipation of World War II. It served as bomber training base and a place of coming and going during the war, as troops transferred to and from the great battlefields of Europe and the Pacific. After the war, it played a role as a staging point in the Berlin Airlift and as the headquarters of the Military Air Transport Service system until 1955. This aerial shows the base in 1957. (MAHS.)

The City of Plymouth commissioned a study to determine the feasibility of constructing an airport. In 1934, Edward Griffith cleared an old apple orchard on South Meadow Road, and on May 17, Alto Sherman landed his biplane on the dirt-and-sand runway. By 1936, Mayflower Airlines began service to Boston, but as a victim of the Depression, the airline soon faded away. (MAHS.)

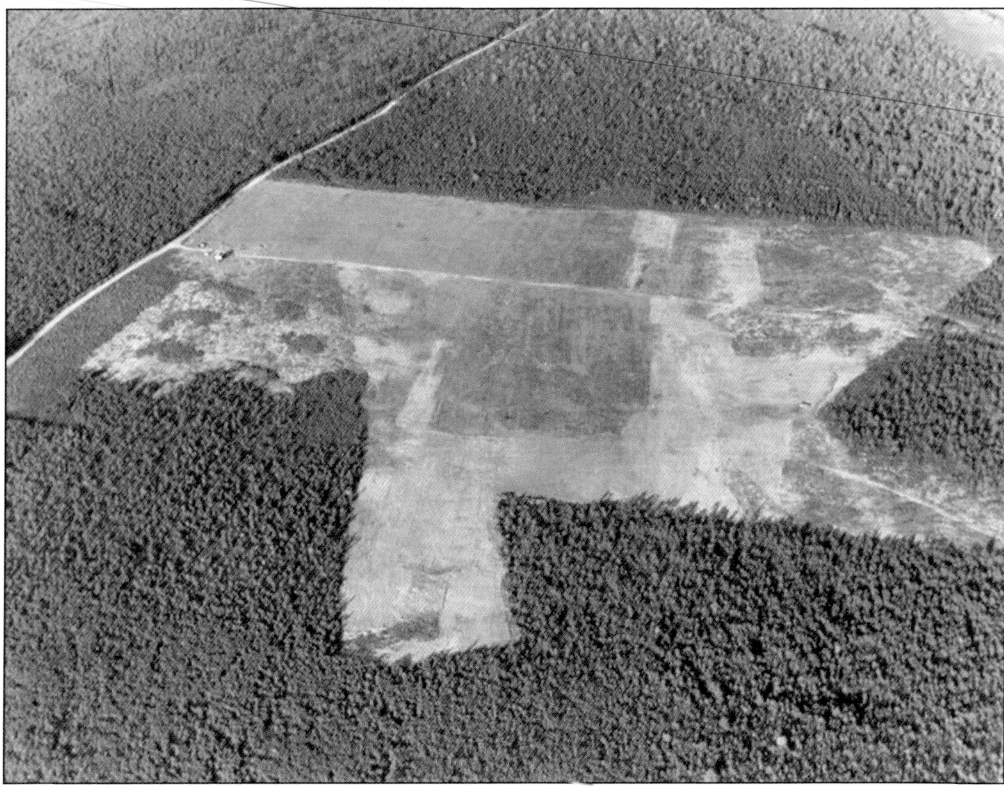

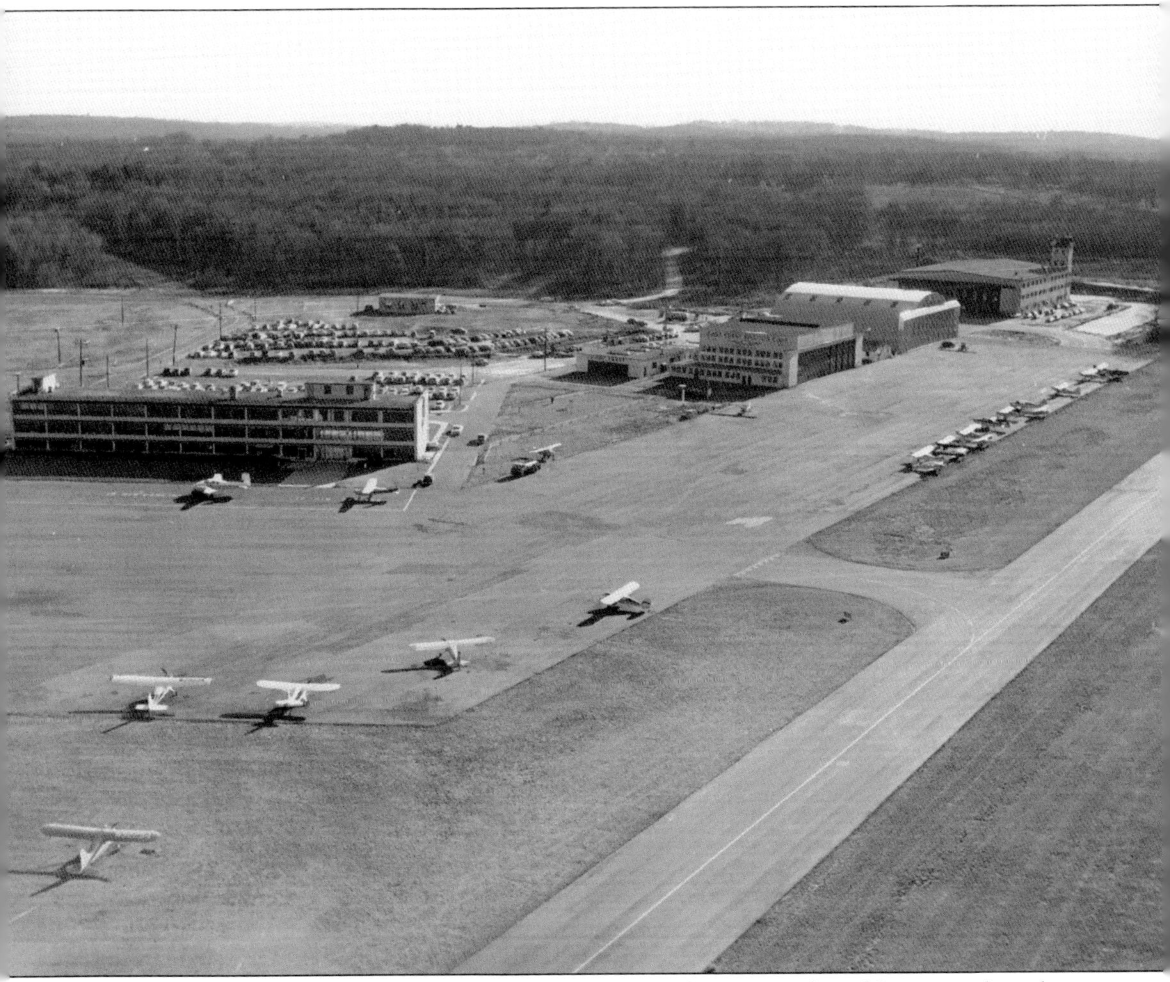

In 1942, the state leased the airport in Bedford (and Lincoln, Concord, and Lexington) to the War Department. There, the Army Air Force trained fighter squadrons to fly off to combat. During these days, too, flyers tested the Massachusetts Institute of Technology's Rad Lab's newest invention: radar. In 1945, Hanscom became the center for the development of electronic systems for the Army Air Force. (MAHS.)

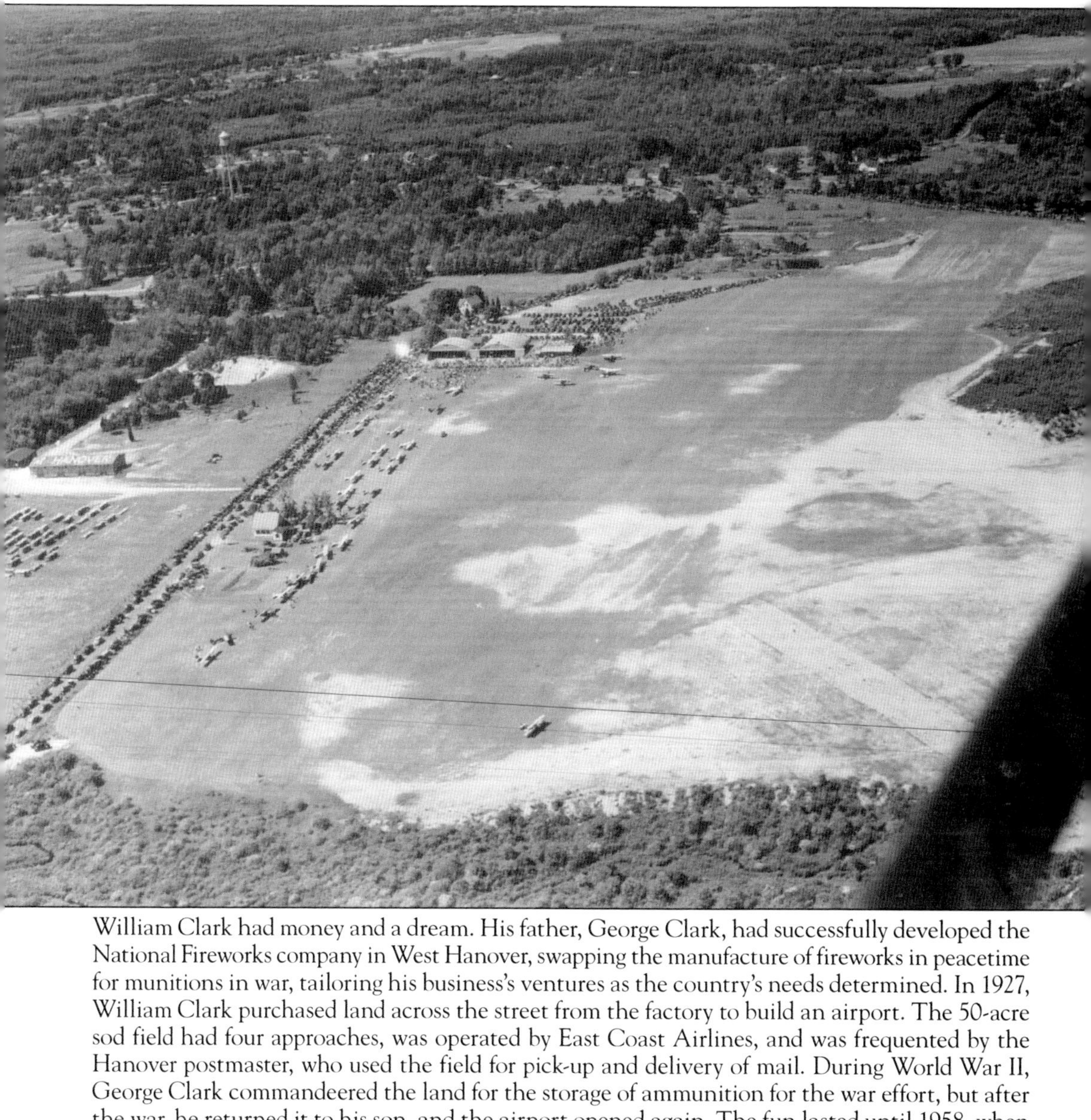

William Clark had money and a dream. His father, George Clark, had successfully developed the National Fireworks company in West Hanover, swapping the manufacture of fireworks in peacetime for munitions in war, tailoring his business's ventures as the country's needs determined. In 1927, William Clark purchased land across the street from the factory to build an airport. The 50-acre sod field had four approaches, was operated by East Coast Airlines, and was frequented by the Hanover postmaster, who used the field for pick-up and delivery of mail. During World War II, George Clark commandeered the land for the storage of ammunition for the war effort, but after the war, he returned it to his son, and the airport opened again. The fun lasted until 1958, when William Clark sold the land to a developer who built residential homes upon it. Today, aerial views and satellite images show no trace of the old field. One hangar now stands in Plymouth, another at the Cranland Airport in Hanson. (MAHS.)

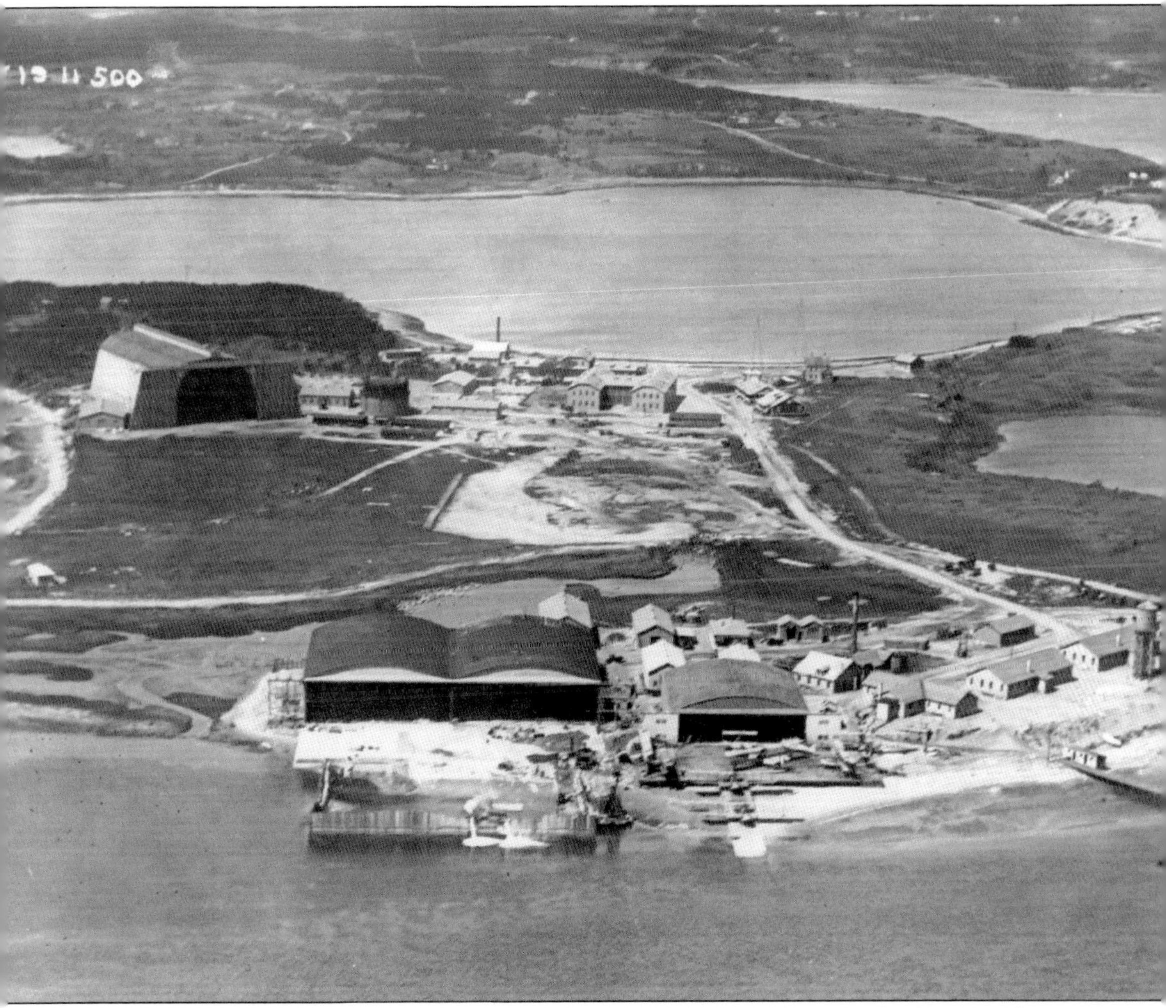

Chatham, at the elbow of the bended arm of Cape Cod, long played a crucial role in anything to do with the approach of ships or planes. With a long marine lifesaving tradition already established before the merging of the Revenue Cutter Service and the Life-Saving Service into the Coast Guard in 1915, the locals' eyes turned toward the sky. The US Navy operated a very short-lived air station out of Nickerson's Neck, Chatham, from 1917 to 1922. Plans were in place and construction started well before America's entry into World War I. Thus, when the time came, the base was ready. A U-boat attack on Nauset Beach scrambled the Chatham pilots in nine Curtiss HS-2L flying boats with plans to bomb the submarine. The U-boat slipped beneath the surface and could not be found, and the shells landed harmlessly on the beach. The base disappeared with little fanfare and is now marked only by a single memorial stone at the end of Strong Island Road. (MAHS.)

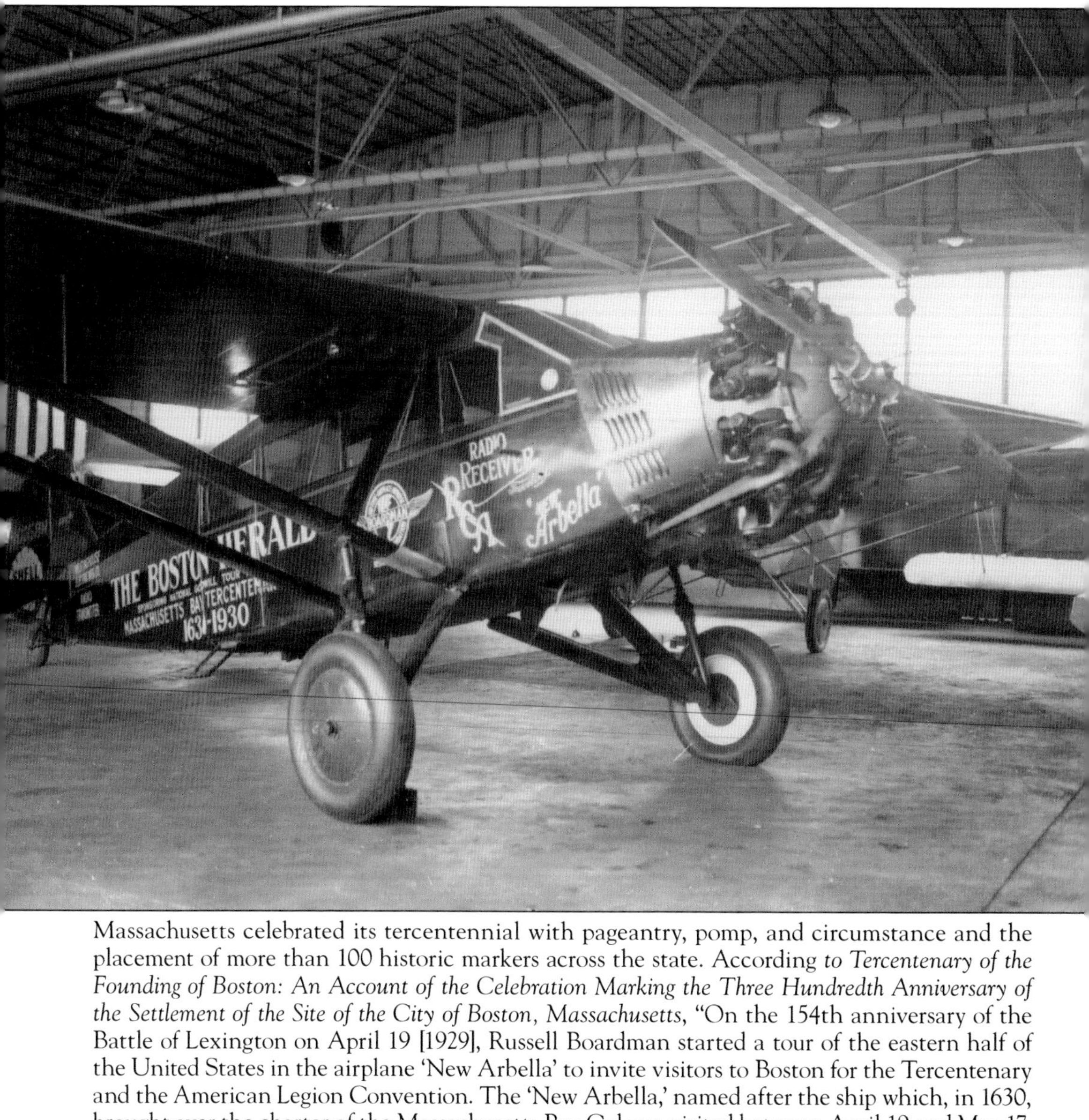

Massachusetts celebrated its tercentennial with pageantry, pomp, and circumstance and the placement of more than 100 historic markers across the state. According *to Tercentenary of the Founding of Boston: An Account of the Celebration Marking the Three Hundredth Anniversary of the Settlement of the Site of the City of Boston, Massachusetts*, "On the 154th anniversary of the Battle of Lexington on April 19 [1929], Russell Boardman started a tour of the eastern half of the United States in the airplane 'New Arbella' to invite visitors to Boston for the Tercentenary and the American Legion Convention. The 'New Arbella,' named after the ship which, in 1630, brought over the charter of the Massachusetts Bay Colony, visited between April 19 and May 17, 65 cities including 21 state capitals. While this flight was in progress, Boston was visited by 67 of the finest fighting planes of the United States Navy, which came on from Hampton Roads and gave a demonstration over the city." (MAHS.)

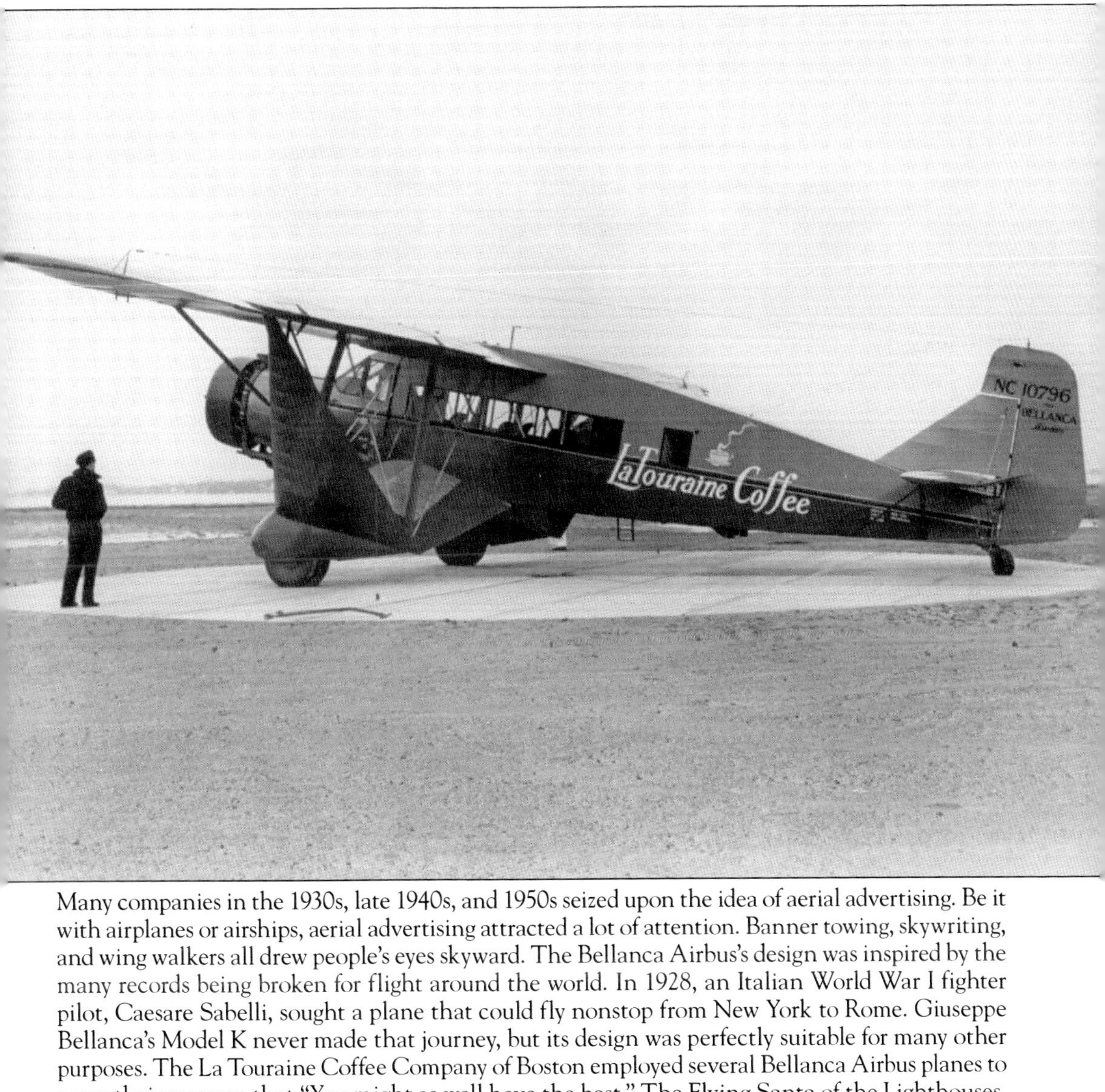

Many companies in the 1930s, late 1940s, and 1950s seized upon the idea of aerial advertising. Be it with airplanes or airships, aerial advertising attracted a lot of attention. Banner towing, skywriting, and wing walkers all drew people's eyes skyward. The Bellanca Airbus's design was inspired by the many records being broken for flight around the world. In 1928, an Italian World War I fighter pilot, Caesare Sabelli, sought a plane that could fly nonstop from New York to Rome. Giuseppe Bellanca's Model K never made that journey, but its design was perfectly suitable for many other purposes. The La Touraine Coffee Company of Boston employed several Bellanca Airbus planes to carry their message that "You might as well have the best." The Flying Santa of the Lighthouses, William Wincapaw, also used Airbuses at Christmas to airdrop gifts to lighthouse keepers and their families, a much appreciated gesture of thanks for a lonely, dangerous job. (MAHS.)

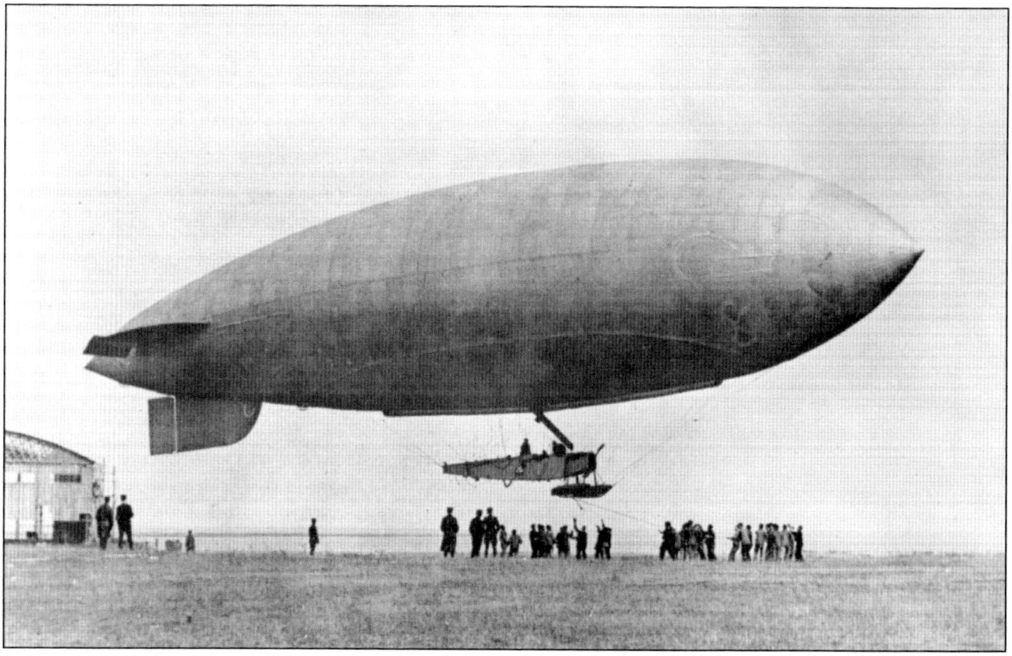

During 1915, the US Navy embarked on a program to use airships as advance scouts for the fleet. Prof. Jerome Hunsaker of MIT, ably assisted by Donald Douglas (later to gain fame with many historic military and commercial aircraft developments), did an incredible amount of the research and development of the Navy's initial airships. The first successful model was the B-class blimp, shown here over Chatham Naval Air Station. (MAHS.)

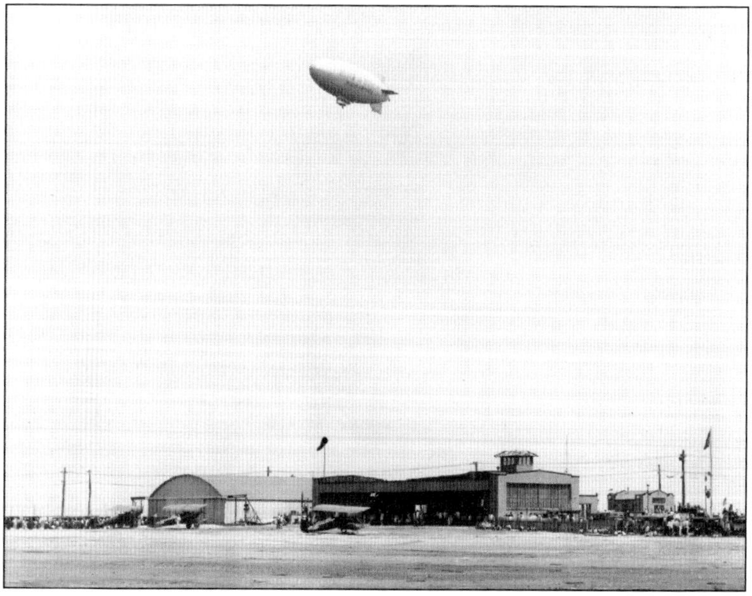

Advertising worked as well on blimps as it did planes. Here, one soars over the Dennison Airport in Quincy. The wording on it says "Bird & Son, Inc., Rugs-Roofs." The company started in Needham in 1795 as a small paper mill on the banks of the Charles River but changed to flooring materials and other products by the 20th century. Obviously, marketing was a strength of the company. (MAHS.)

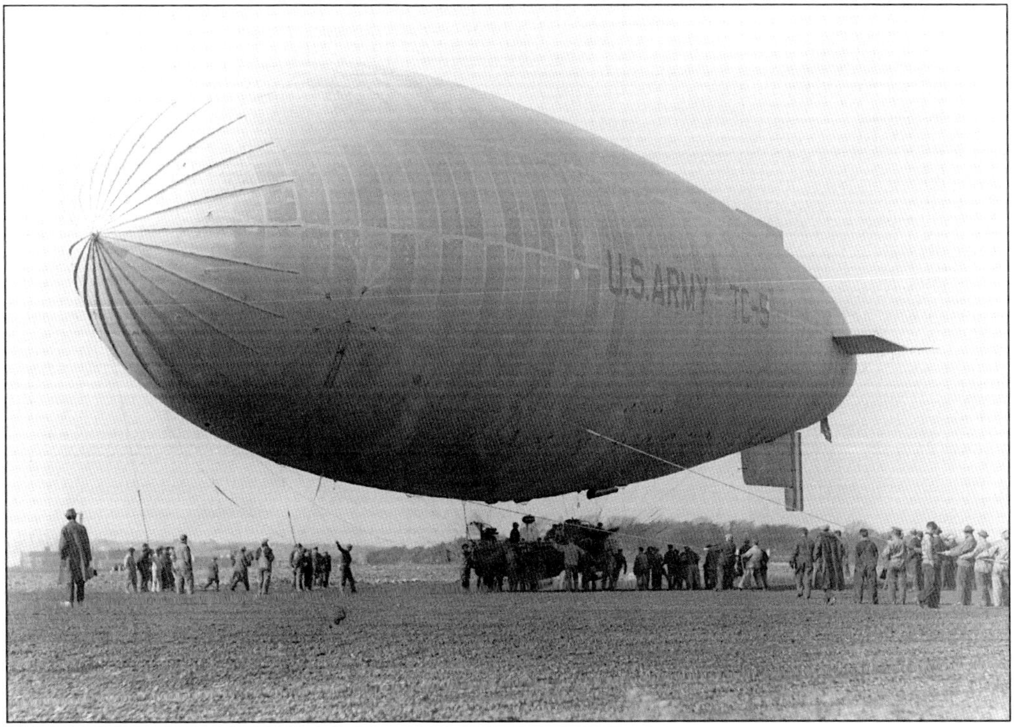

The US Navy was not the only branch of the military to see the value in LTA vehicles for defense and advanced scouting purposes. The TC-5 was the primary training (T), Type C airship for the US Army. While never stationed in Massachusetts, the TC-5 did spend a lot of time operating off the East Coast. The TC-5 is seen here in October 1926. (MAHS.)

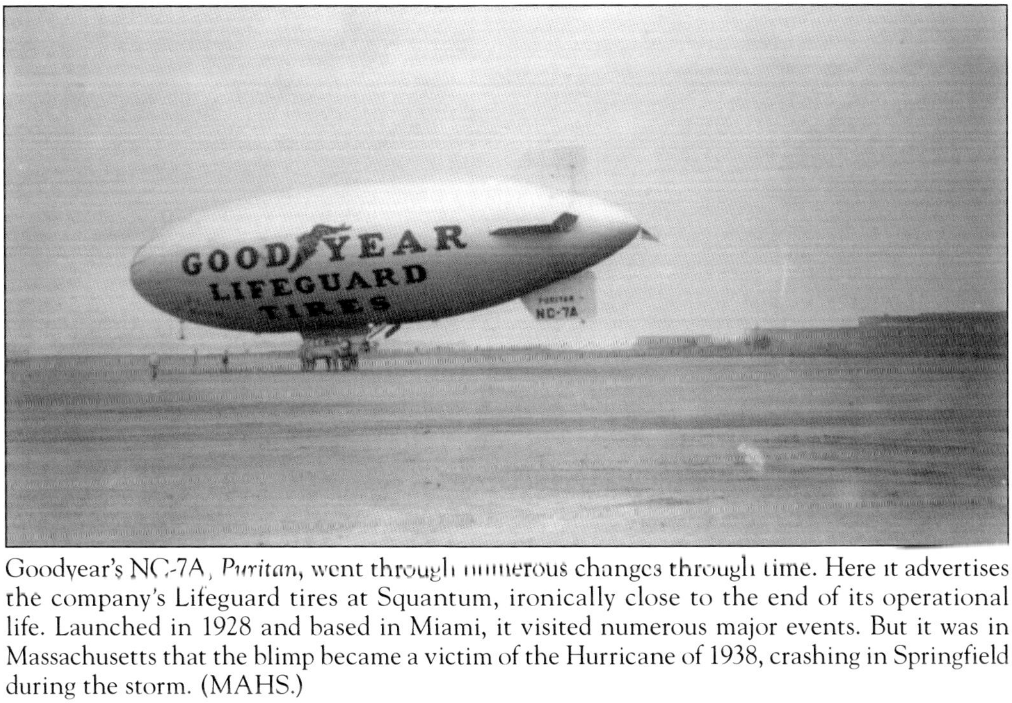

Goodyear's NC-7A, *Puritan*, went through numerous changes through time. Here it advertises the company's Lifeguard tires at Squantum, ironically close to the end of its operational life. Launched in 1928 and based in Miami, it visited numerous major events. But it was in Massachusetts that the blimp became a victim of the Hurricane of 1938, crashing in Springfield during the storm. (MAHS.)

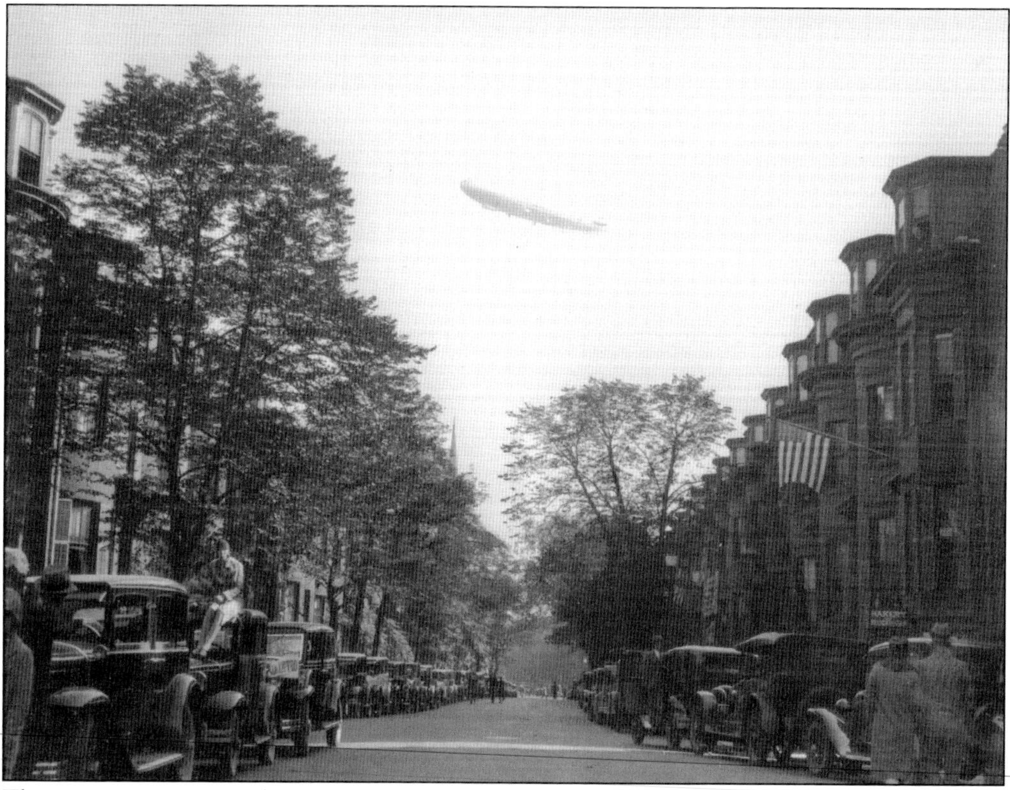

The preeminent LTA airship company in the world manufactured them in Germany under Count Otto von Zeppelin. Zeppelin's dirigibles traveled throughout Europe, the United States, South America, the Far East, and around the world on a somewhat regular schedule. The most famous Zeppelin was (and still is) the *Hindenburg.* (MAHS.)

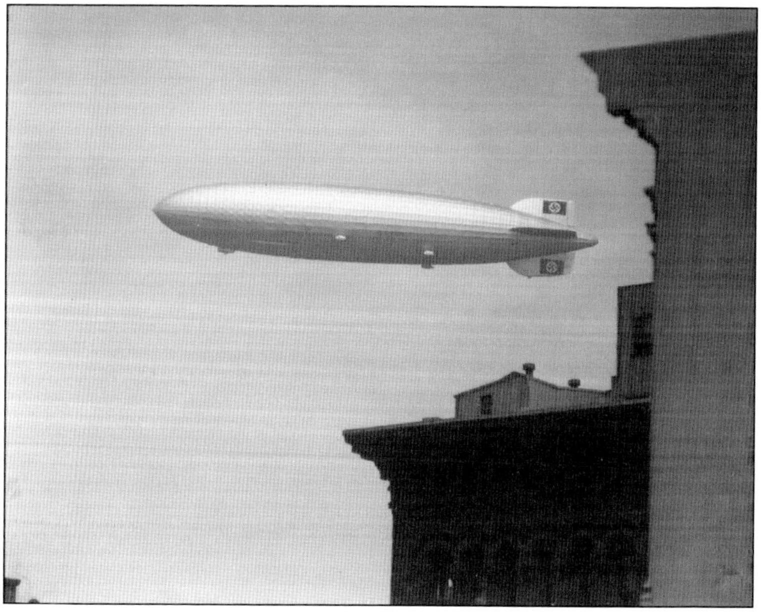

Hindenburg reached Boston on May 6, 1937, on its way to Lakehurst, New Jersey. Named for Field Marshal Paul von Hindenburg, president of Germany from 1925 to 1934, the rigid airship was a point of pride for the nation rebuilding after World War I and a symbol of German ingenuity. (MAHS.)

Spectators filled the streets and hung out of windows in South Boston as the airship passed over the city that day. The Nazi swastikas emblazoned on *Hindenburg* did not yet have the meaning they would in just a few short years, symbolic of Adolf Hitler and his attempt to take over Europe. (MAHS.)

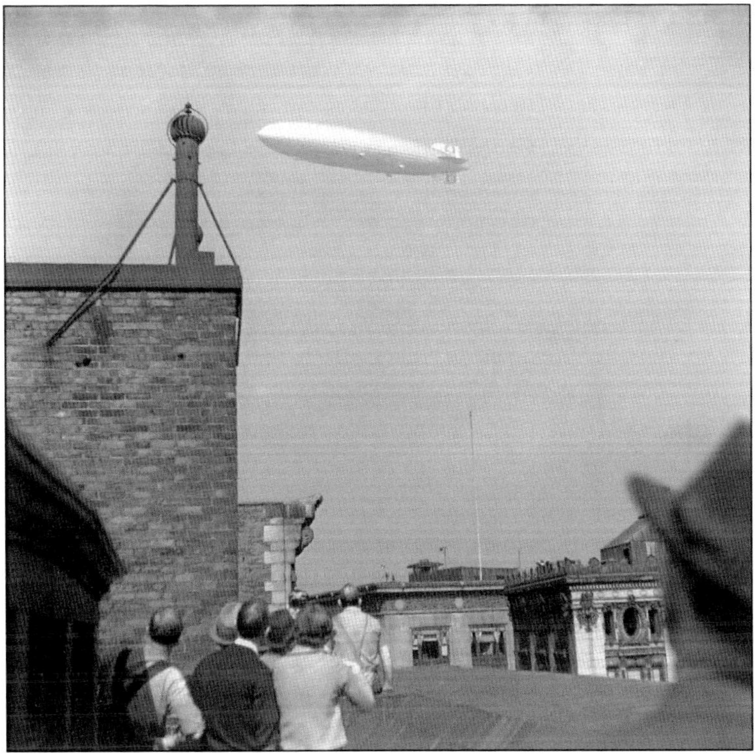

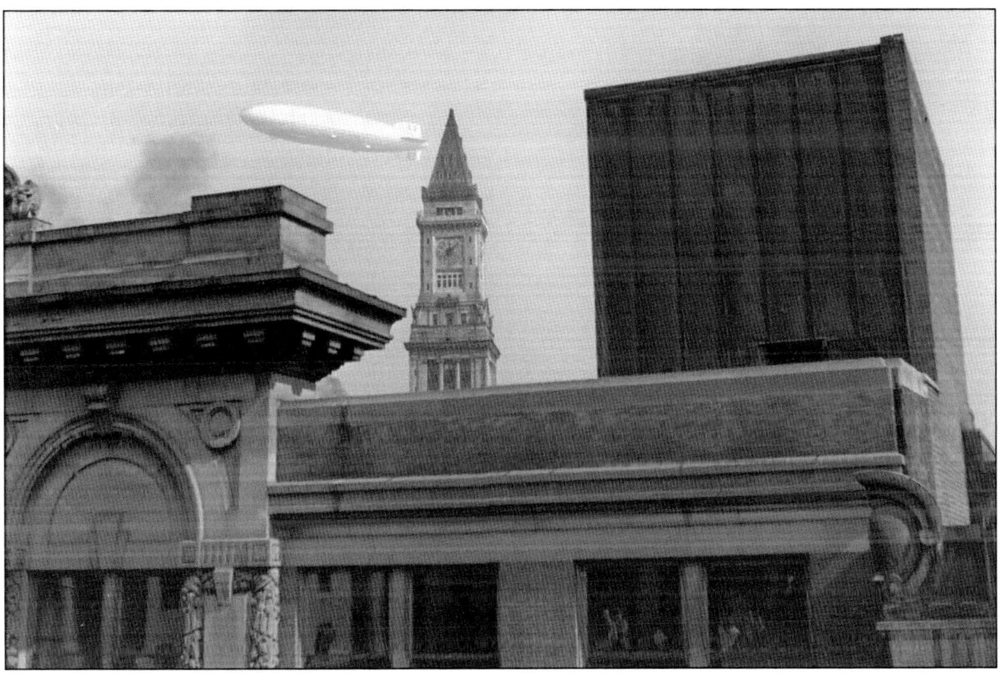

Passing through the city (here, with the customs house tower in the background), *Hindenburg* flew south over Manhattan and onto Lakehurst, where it crashed in a fiery disaster, killing 35 of the 97 people aboard as well as one worker on the ground. The residents of South Boston who watched it pass over the city were among the last people to see *Hindenburg* before the disaster. (MAHS.)

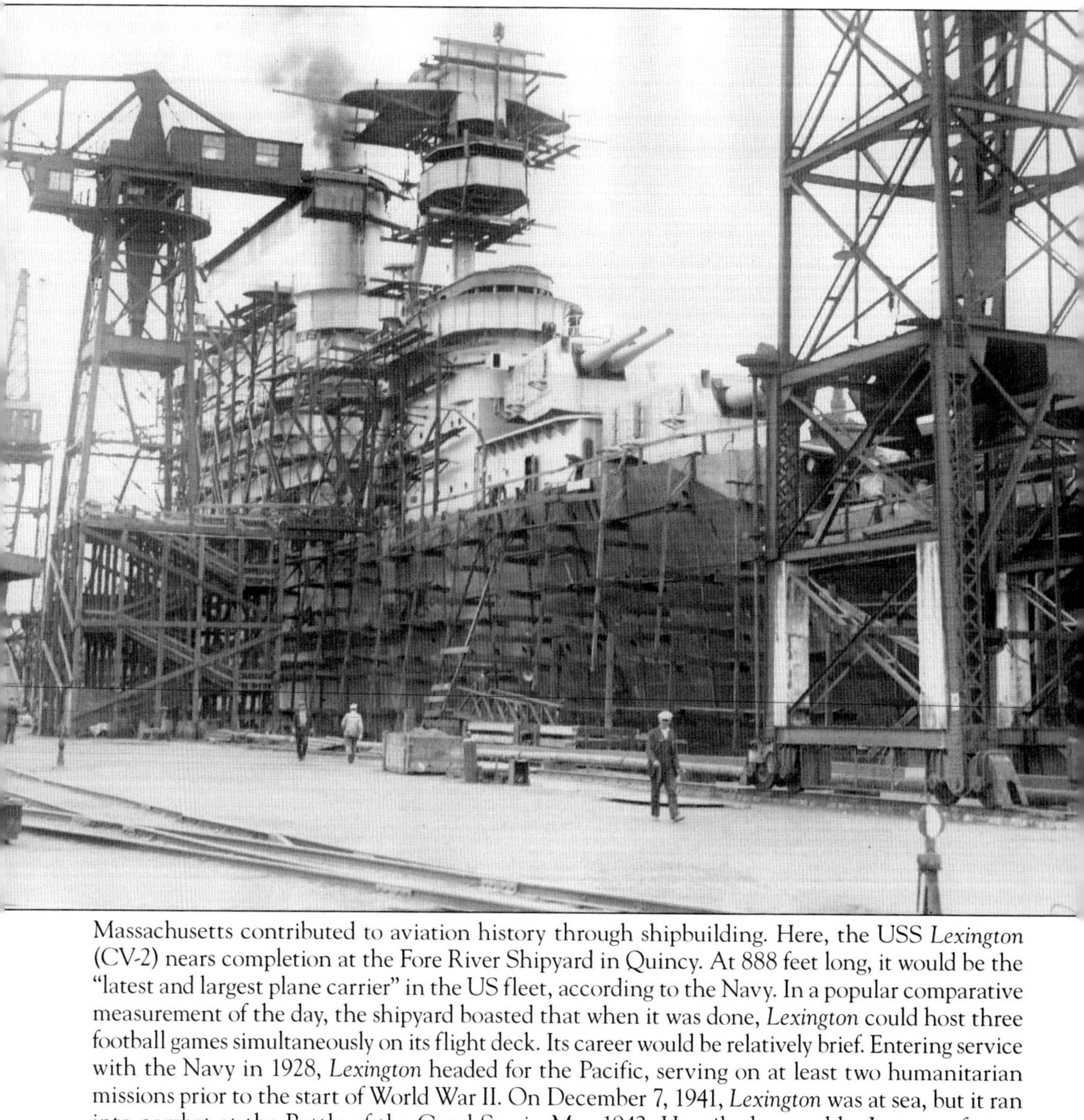

Massachusetts contributed to aviation history through shipbuilding. Here, the USS *Lexington* (CV-2) nears completion at the Fore River Shipyard in Quincy. At 888 feet long, it would be the "latest and largest plane carrier" in the US fleet, according to the Navy. In a popular comparative measurement of the day, the shipyard boasted that when it was done, *Lexington* could host three football games simultaneously on its flight deck. Its career would be relatively brief. Entering service with the Navy in 1928, *Lexington* headed for the Pacific, serving on at least two humanitarian missions prior to the start of World War II. On December 7, 1941, *Lexington* was at sea, but it ran into combat at the Battle of the Coral Sea in May 1942. Heavily damaged by Japanese forces, *Lexington* had to be sunk by friendly fire on May 8 so that it would not be taken by the enemy. Then building a new carrier, shipyard workers at Fore River called Washington, DC, and suggested a name change from *Cabot* to *Lexington*, which was approved for the new vessel. (MAHS.)

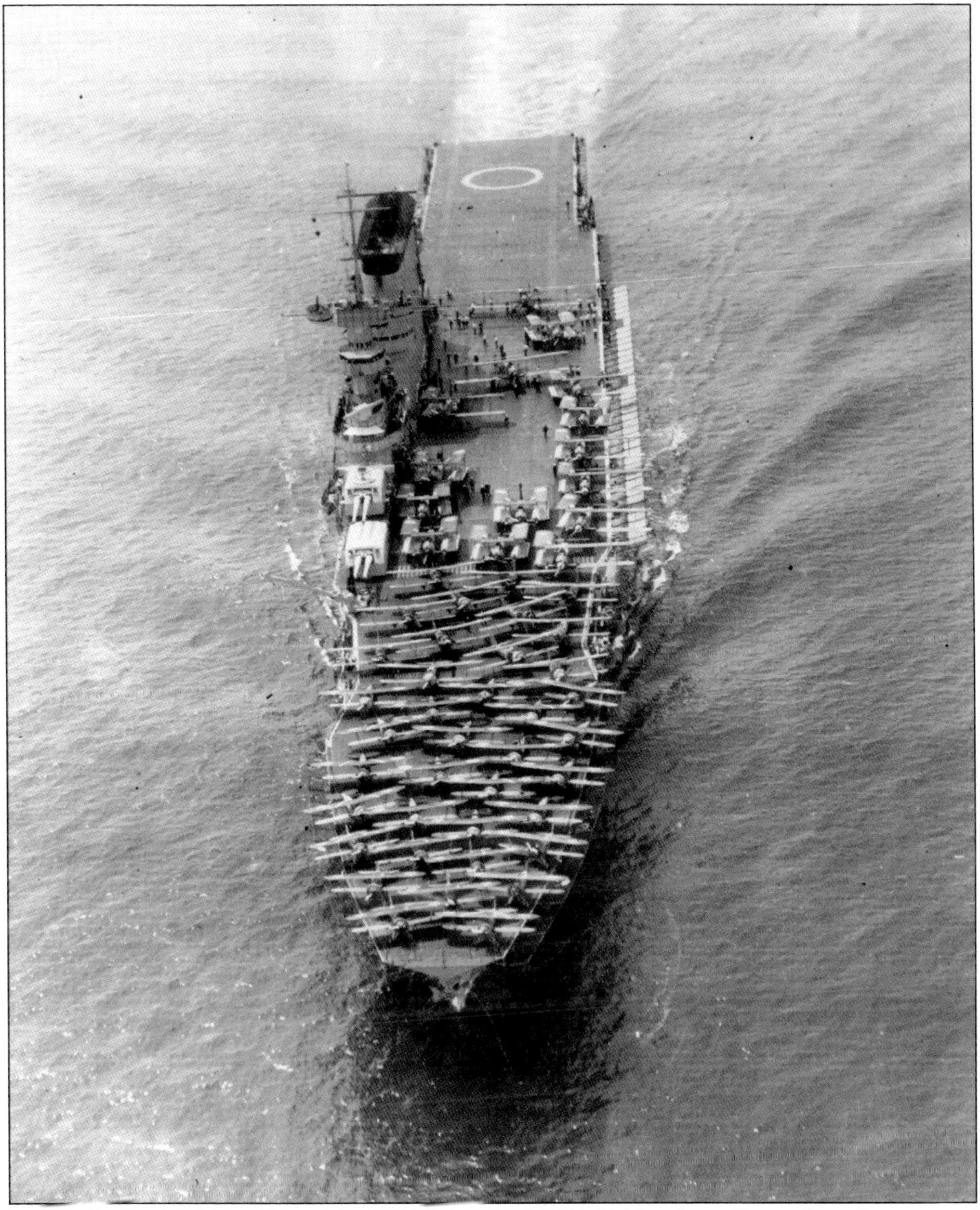

Originally, *Lexington* could carry 78 aircraft, which included 36 bombers. Shown here during naval maneuvers off the Virginia coast, it impressed the people it needed to most. "While congressmen and officials watched interestedly," read an official press clipping, "these tiny corsair planes quickly laid a protecting smoke screen around the fleet." In 1936, the planes loaded onto her deck included 18 Grumman F2F-1 fighter planes, the Navy's standard fighter from 1936 to 1940, and 18 Boeing F4B-4 fighter planes, with nine more F2Fs held in reserve. Dive bombers and torpedo bombers added extra wallop, while Grumman Ducks (amphibious planes, which could land on water or land) and Vought O2U Corsair observation planes filled out the complement, with many others in reserve, bringing its mid-Depression air group to a total of more than 100 planes. Among *Lexington*'s many missions was the search for the missing Amelia Earhart in 1937. (MAHS.)

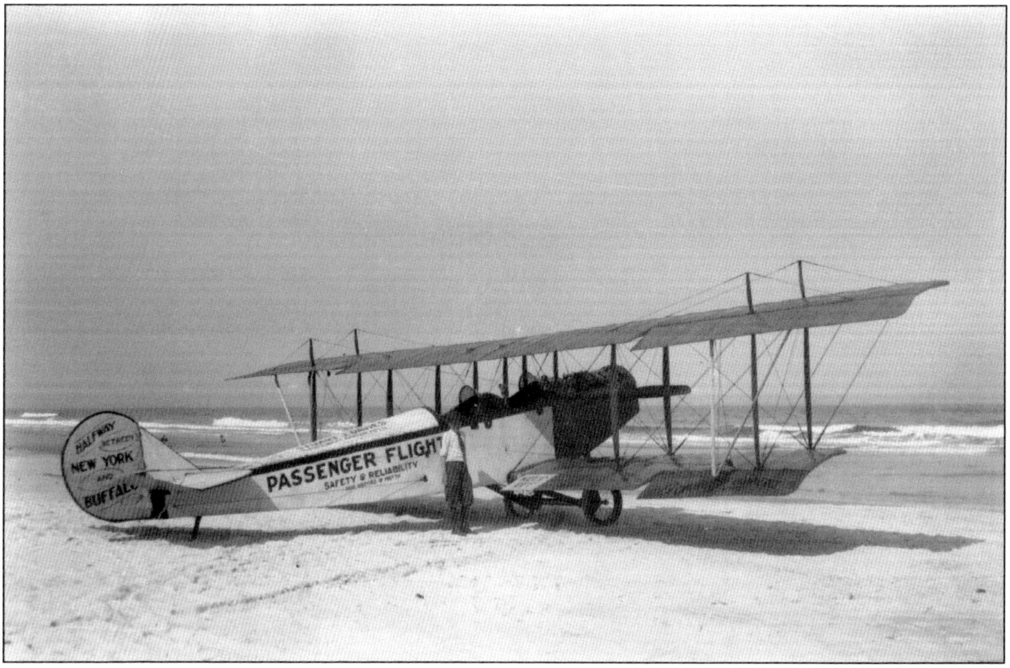

Stuart L. Davies is shown here with his favorite plane off Squantum. Davies flew out of Glenn D. Wicks Field in Utica, New York, hence the advertisement on the tail about being halfway between New York and Buffalo. Wicks had volunteered for Army air service in World War I, collided with another pilot, and crashed and died behind German lines. Utica's airfield was posthumously named in his honor. (MAHS.)

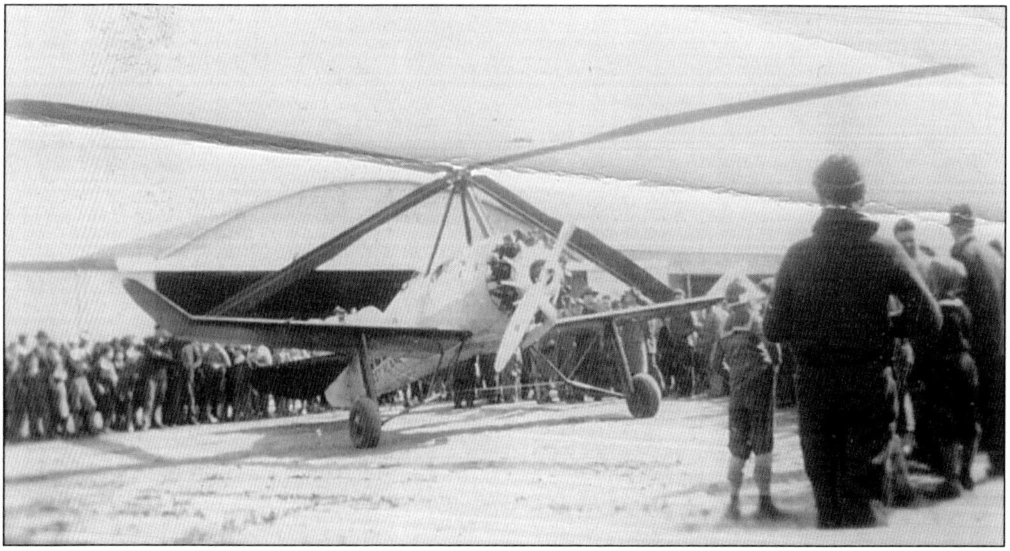

Juan de la Cierva's autogyro played an important role in aviation history, allowing for vertical takeoffs long before the development of the helicopter. For a while, the record for height in an autogyro belonged to Amelia Earhart at 18,415 feet, set at Pitcairn Field in Pennsylvania. Earhart toured the country doing demonstrations of the autogyro's capabilities, including a stop at Squantum in Quincy. (QHS.)

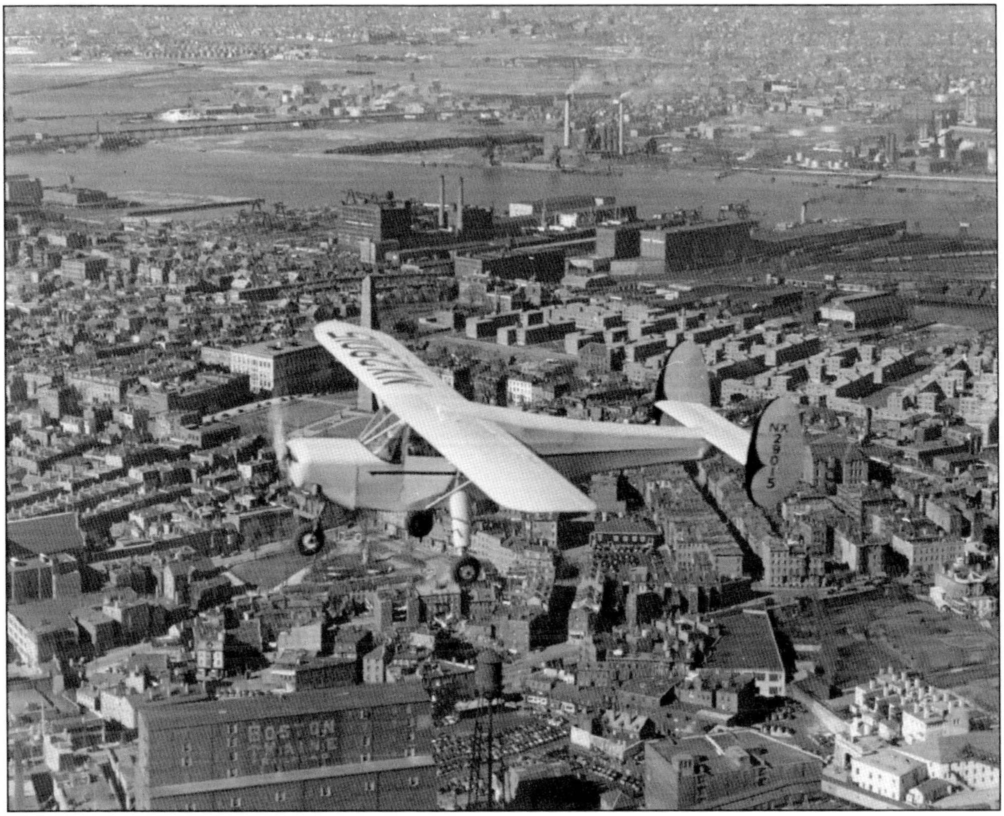

Otto Koppen graduated from the Massachusetts Institute of Technology, then returned a few years later to teach stability and control as part of his focus in aeronautical engineering. One of the planes he designed, the G1-80 Skyfarer, was built by the General Aircraft Corporation of Lowell, seen here flying over Boston. (MAHS.)

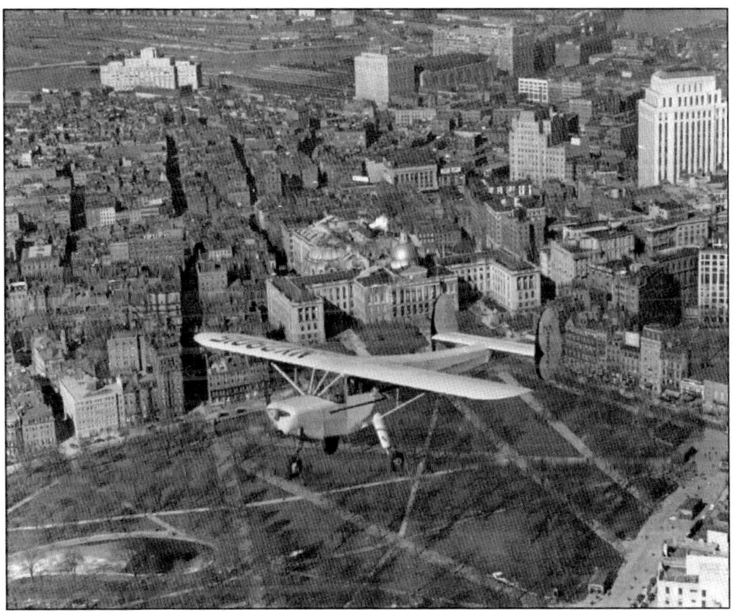

The timing for the Skyfarer could not have been worse. Although the company stated in *Popular Science* in 1941 that the plane was so simple to fly that an average person could learn how to do so in about an hour, its date with destiny was cut short by the coming of World War II. General Aircraft Corporation built 17, then sold the manufacturing rights to a Great Lakes concern that built just two more. (MAHS.)

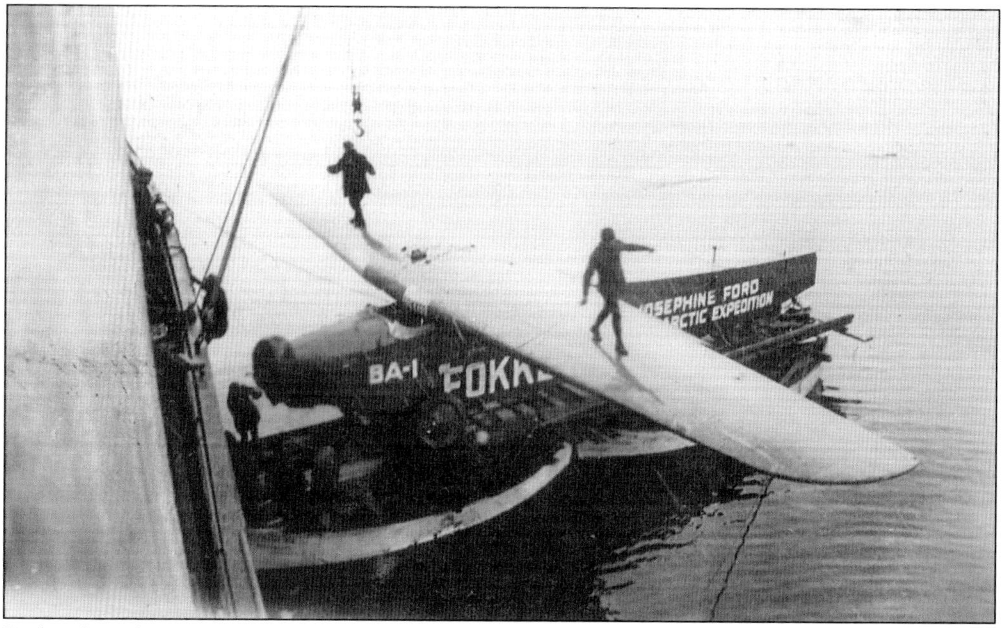

Richard Byrd, at numerous points in his Navy career, called Boston home. Among his many goals in life was to be the first person to fly over the North Pole, and thanks to the patronage of Edsel Ford, president of the Ford Motor Company, he did so. Flying a plane named for Ford's daughter Josephine, Byrd and Navy chief aviation pilot Floyd Bennett flew this Fokker VIIa/3m on what is still considered a disputed flight today. (MAHS.)

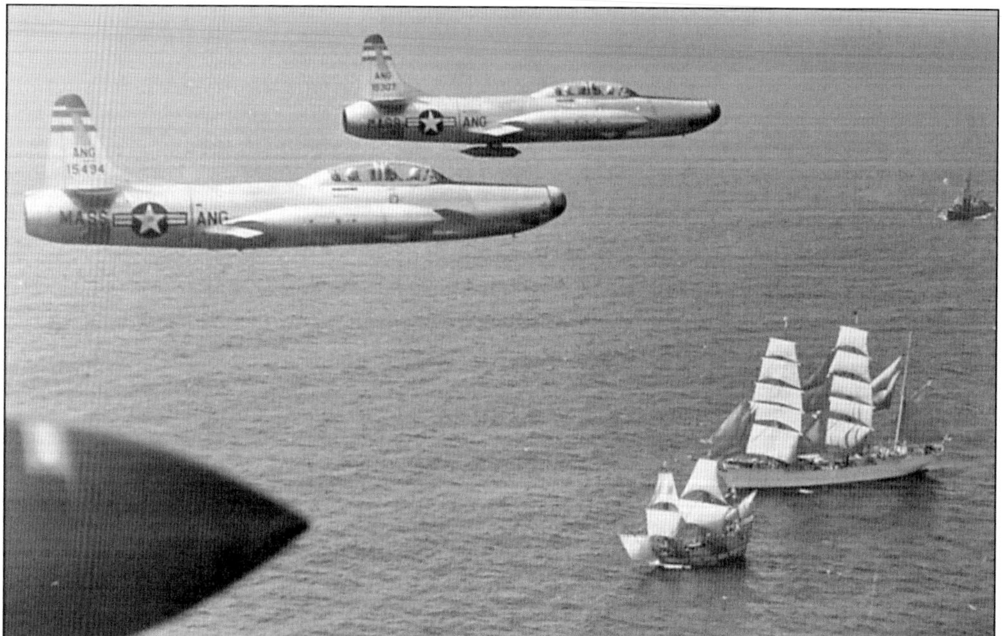

When the *Mayflower II* finally reached the Massachusetts coast, the replica was greeted by throngs on the Plymouth coastline, dozens of boaters, and a pair of Massachusetts Air National Guard F-94 Starfires. The National Guard accepted them in 1951, replacing its F-86A Sabres. This scene took place six years later. (MAHS.)

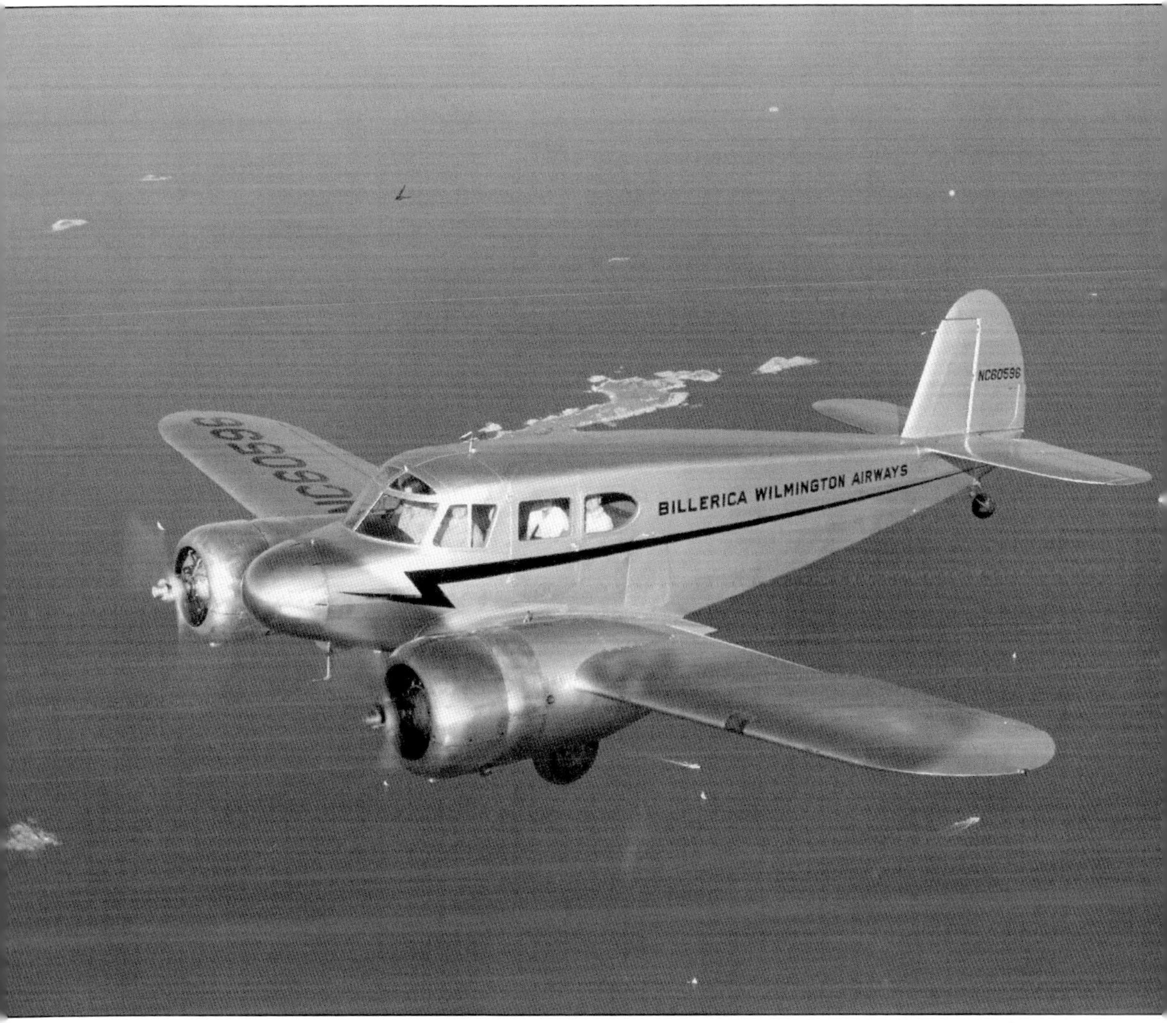

Russell Todman started the Billerica-Wilmington airfield, or Shawsheen Pines Airport, but it was once Fred Cain joined him that things really took off. Cain, a major dealer in International Harvester heavy equipment, had the tools to turn the airfield into an airport. Cain flew to Texas to collect a few AT-6 Texans, hired a flight instructor and a mechanic, and opened a flight school. Here, a Cessna Model T-50 bearing an undeniable identifying label flies northwest over Marblehead, with Cat Island in the background. An ad in the March 1947 edition of *Flying* magazine stated, "Shawsheen has a spacious and well-equipped repair shop . . . the 2300-foot lighted macadam runway will soon be augmented by two others of even greater length." The airfield was located off Alexander Road, but by the early 1960s, it had gone out of business. Aerial imagery today shows no trace of the old airfield, as, like most of Massachusetts's farmland, the open space has reforested. (MAHS.)

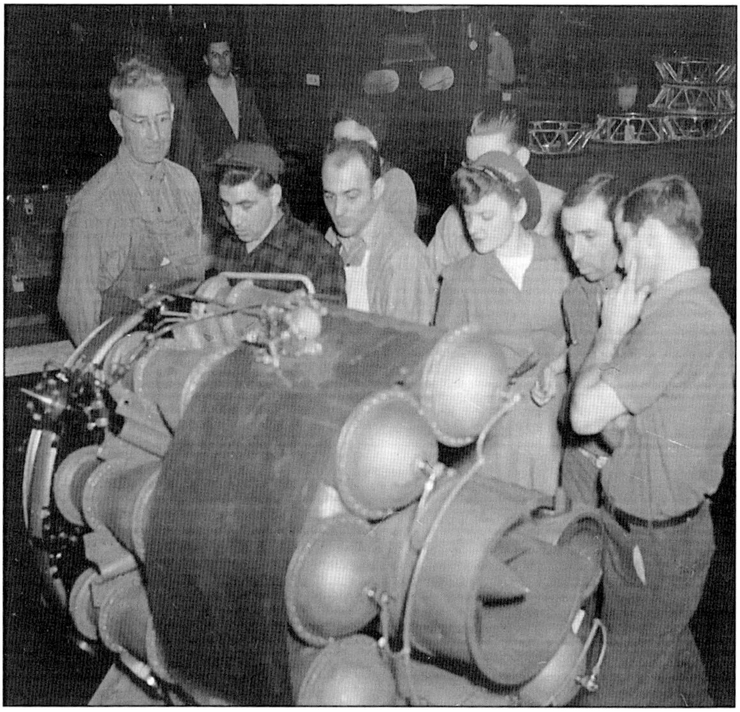

General Electric played several important early roles in the history of aviation. First, in 1917, the federal government sought a manufacturer that could produce a turbo-supercharger to boost airplane performance at high altitudes. GE accepted the challenge first, but when a second company also picked up the gauntlet, the first aircraft engine competition was underway. (David Carpenter/GE.)

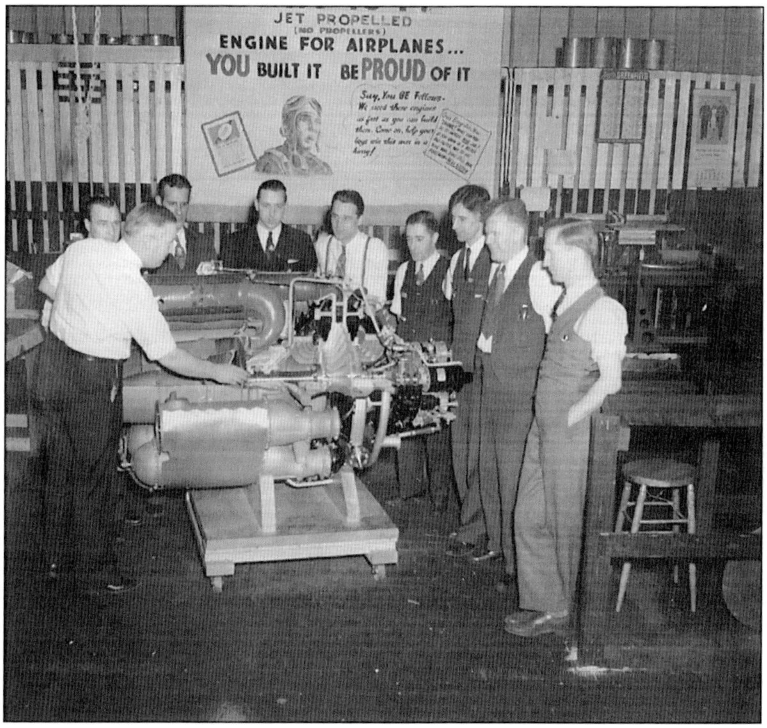

GE won that competition, awarded a contract that set it on a path to achieve several aviation technology firsts. In 1941, the US Army approached GE and asked whether or not the company could replicate or possibly improve the jet propulsion engine designed by Sir Frank Whittle of Great Britain. The result, shown in this series of images, took just six months, unveiled on April 18, 1942. (David Carpenter/GE.)

Six months later, the engine, the I-A, was tested at Muroc Dry Lake, California. Until that time, every American plane had utilized propellers to power its flight. The day after the successful testing of the I-A, Maj. Gen. Henry "Hap" Arnold called Lawrence Dale Bell, head of Bell Aircraft Corporation, to design a plane on which to test it. (MAHS.)

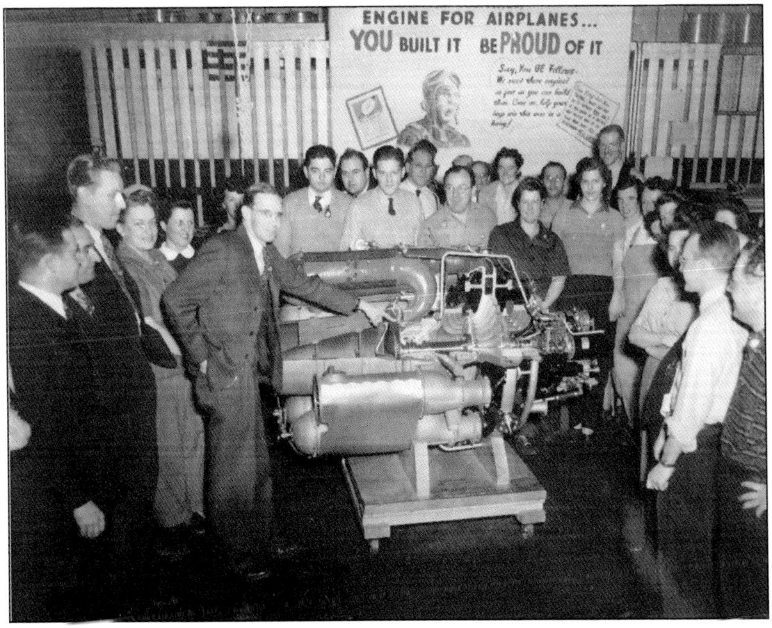

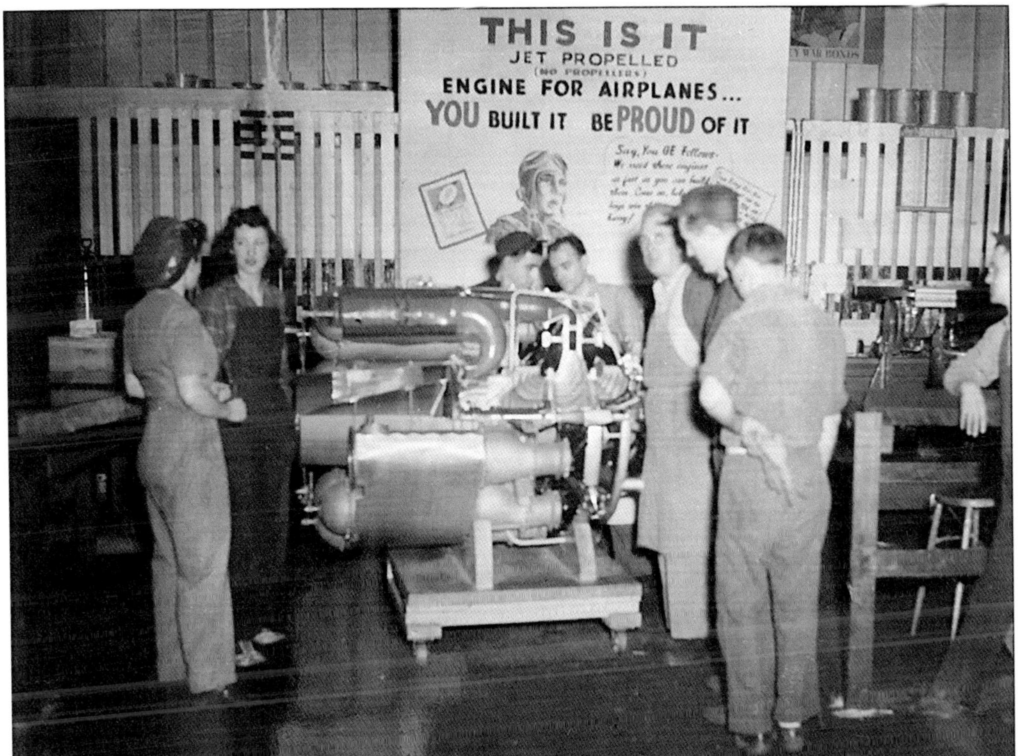

Bell designed the XP-59A Airacomet for the October 1, 1942, test. To disguise the true nature of the plane as it transited to the testing site, the Army Air Force fitted it with a dummy propeller. But as the Airacomet took to the sky that day, powered by two General Electric I-A engines, there could be no doubt that America had soared into the jet age. (David Carpenter/GE.)

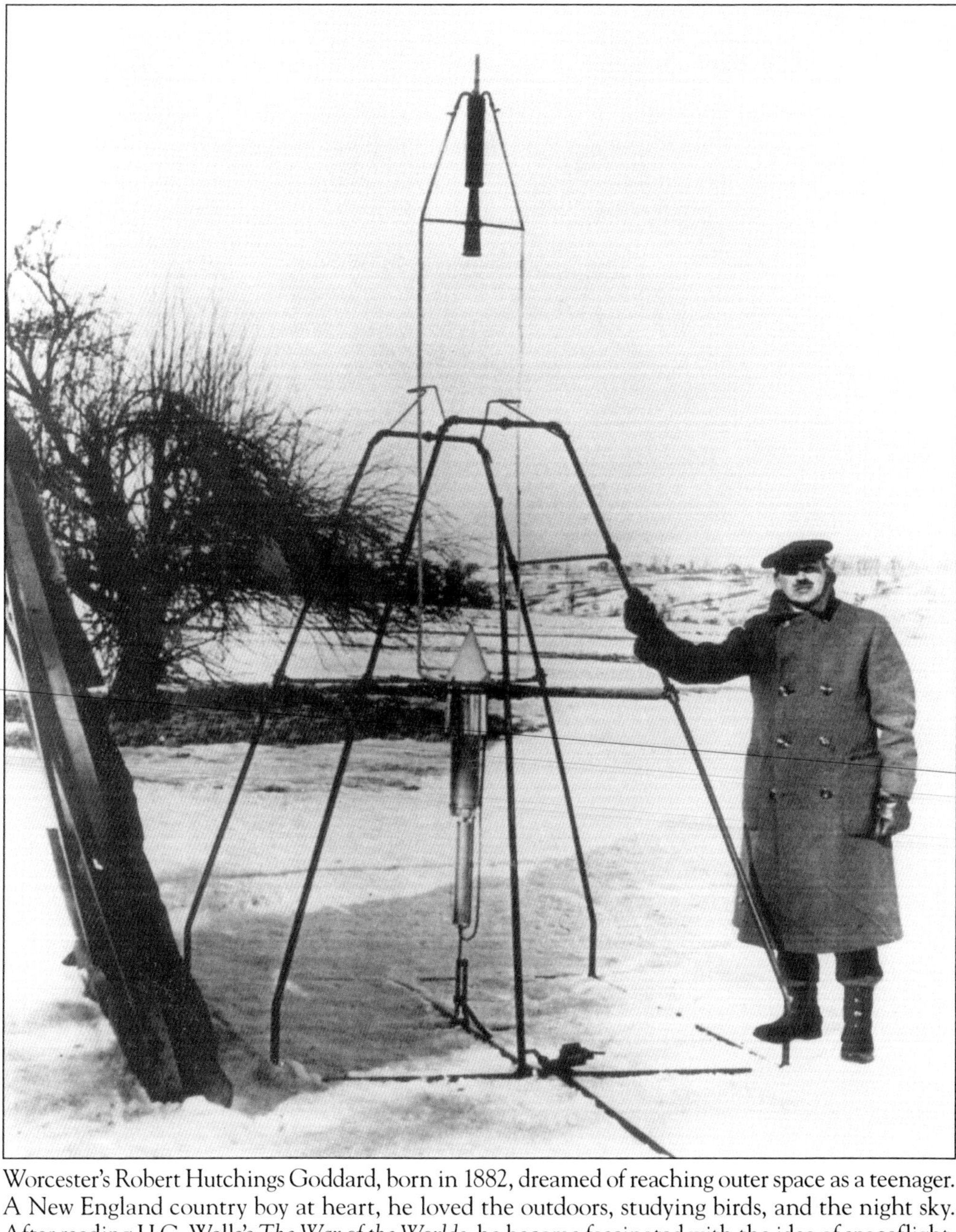

Worcester's Robert Hutchings Goddard, born in 1882, dreamed of reaching outer space as a teenager. A New England country boy at heart, he loved the outdoors, studying birds, and the night sky. After reading H.G. Wells's *The War of the Worlds*, he became fascinated with the idea of spaceflight. He dated this inspiration to October 19, 1899, which he privately called "Anniversary Day" for the rest of his life. Studying at Worcester Polytechnic Institute and Clark University, he earned his doctorate in physics, then accepted a research fellowship at the Palmer Physical Laboratory at Princeton University. Two of his 214 patented inventions led to the dawn of the space age: his multistage rocket and his liquid-fueled rocket, the launching frame of which is shown here. He tested his first powder rocket in 1914 at Clark and deduced that atmospheric conditions act detrimentally on propulsion and that in a vacuum like space, a rocket's capabilities are increased. Between 1926 and 1941, Goddard and his students launched more than 30 rockets. (MAHS.)

Founded in 1935 in Worcester, the David Clark Company is an American leader in the design and manufacturing of noise-attenuating headsets. It started in the textile business developing unique knitted materials and advanced to making aerospace- and communications-related products, including aerospace and industrial protective equipment, pressure spacesuit systems, and anti-G suits. Here, a pilot models an advanced lightweight pressure suit in front of a U-2. (MAHS.)

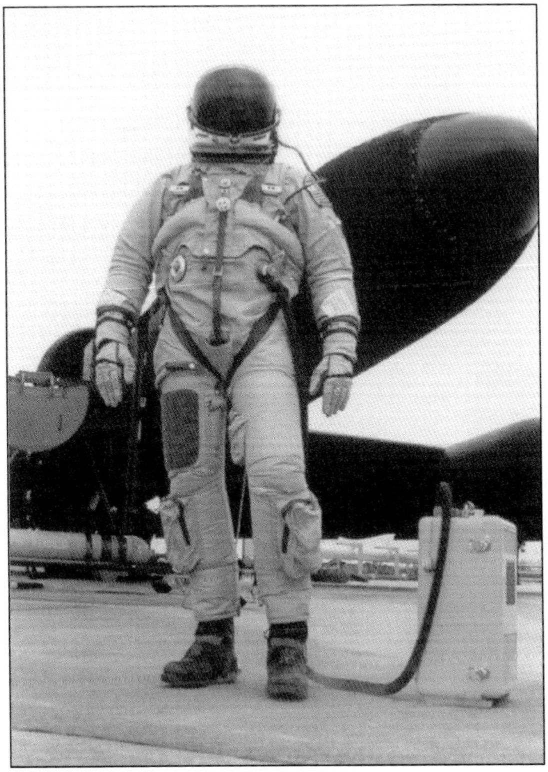

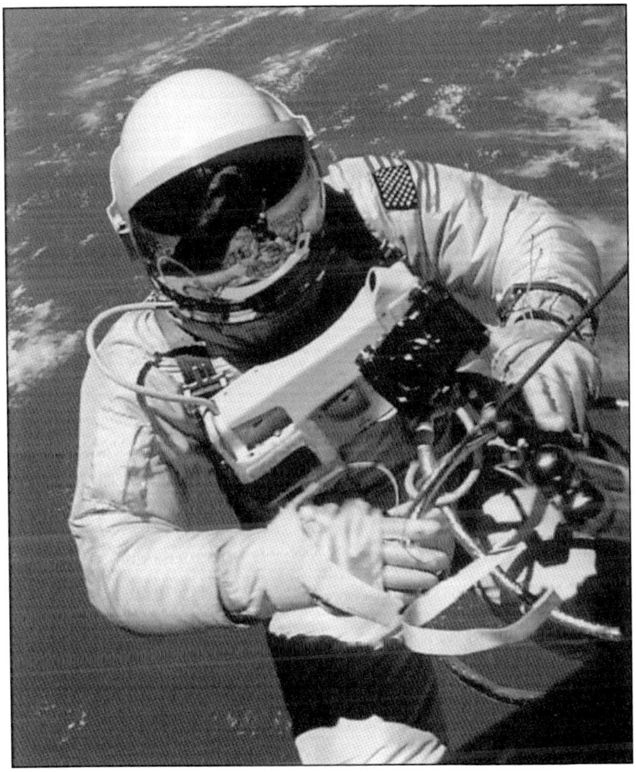

The company has been involved with the design and development of air and space crew protective equipment since 1941 and eventually manufactured all of the spacesuits worn by Gemini crews. The G4C suit, shown here, was worn by all crews from Gemini 4 to Gemini 12, excluding Gemini 7, and was the suit Ed White wore when he made the first spacewalk on June 3, 1965. (MAHS.)

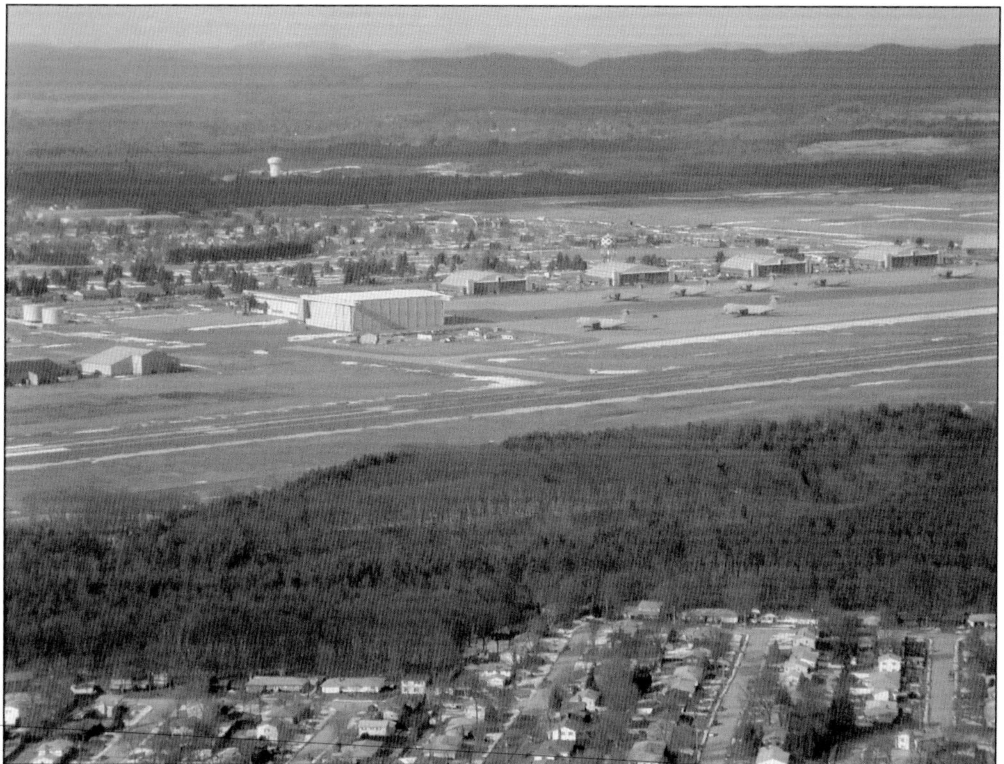

While few of the many airfields founded in Massachusetts during the early days of aviation have survived, some have expanded and thrived. Westover is today the Westover Air Force Reserve Base, the working home of approximately 5,500 military and civilian employees. It is the largest Air Force Reserve base in the United States. (Andrew Biscoe.)

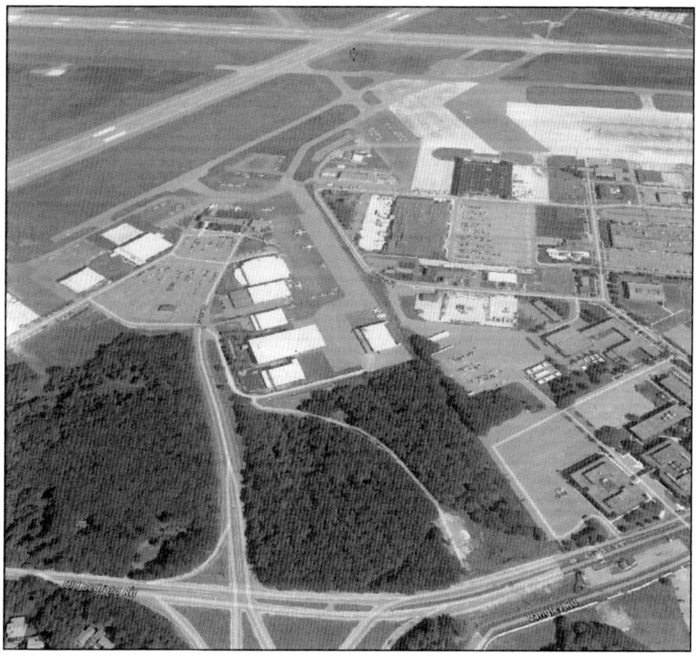

Hanscom today lives a double life. As Providence's T.F. Green and New Hampshire's Manchester Airports have commercially grown, Hanscom has become important as a public general aviation airport. The runways are used, too, by the Hanscom Air Force Base, where research continues today on electronics. The Massachusetts Institute of Technology moved into its new Lincoln Lab in the 1950s. (MAHS.)

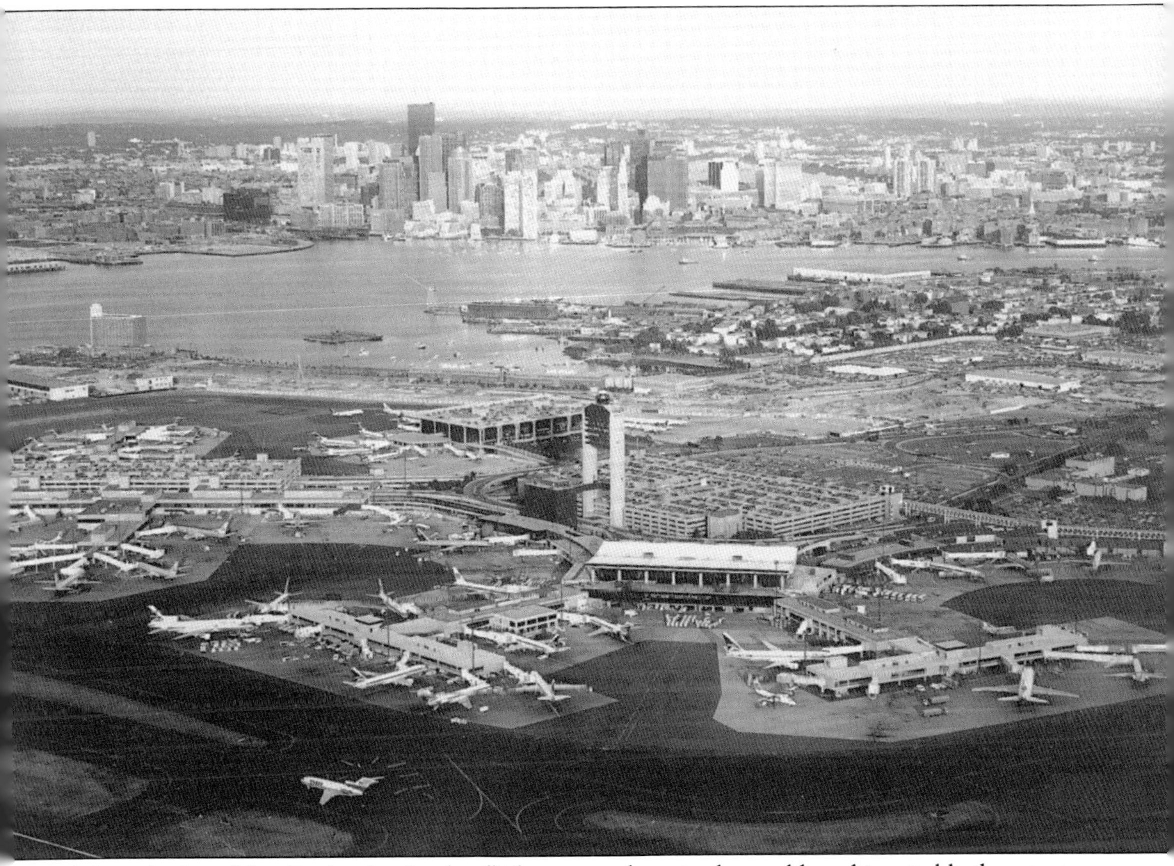

Ask anybody in New England to book a flight to anywhere in the world, and inevitably they start with the airport abbreviation "BOS" for Logan International Airport. The airport has grown in infrastructure and amenities through time. In 1973, the iconic twin-pylon-supported control tower was added, as were two terminals. In 1974, the Bird Island Flats were finally filled, adding 234 acres to the grounds. By 1980, more than 15 million passengers used the airport annually. During the 1980s, Massport focused on noise abatement and even soundproofing. By the 1990s, the average annual usage reached 25 million passengers. Major improvements took place in 2006, including new covered walkways, moving sidewalks connecting all terminals with the central parking garage, a new MBTA Airport station, and a split-level road system separating vehicular traffic between arrivals and departures. Today, Logan Airport stimulates the New England economy to the tune of $7 billion annually. (MAHS.)

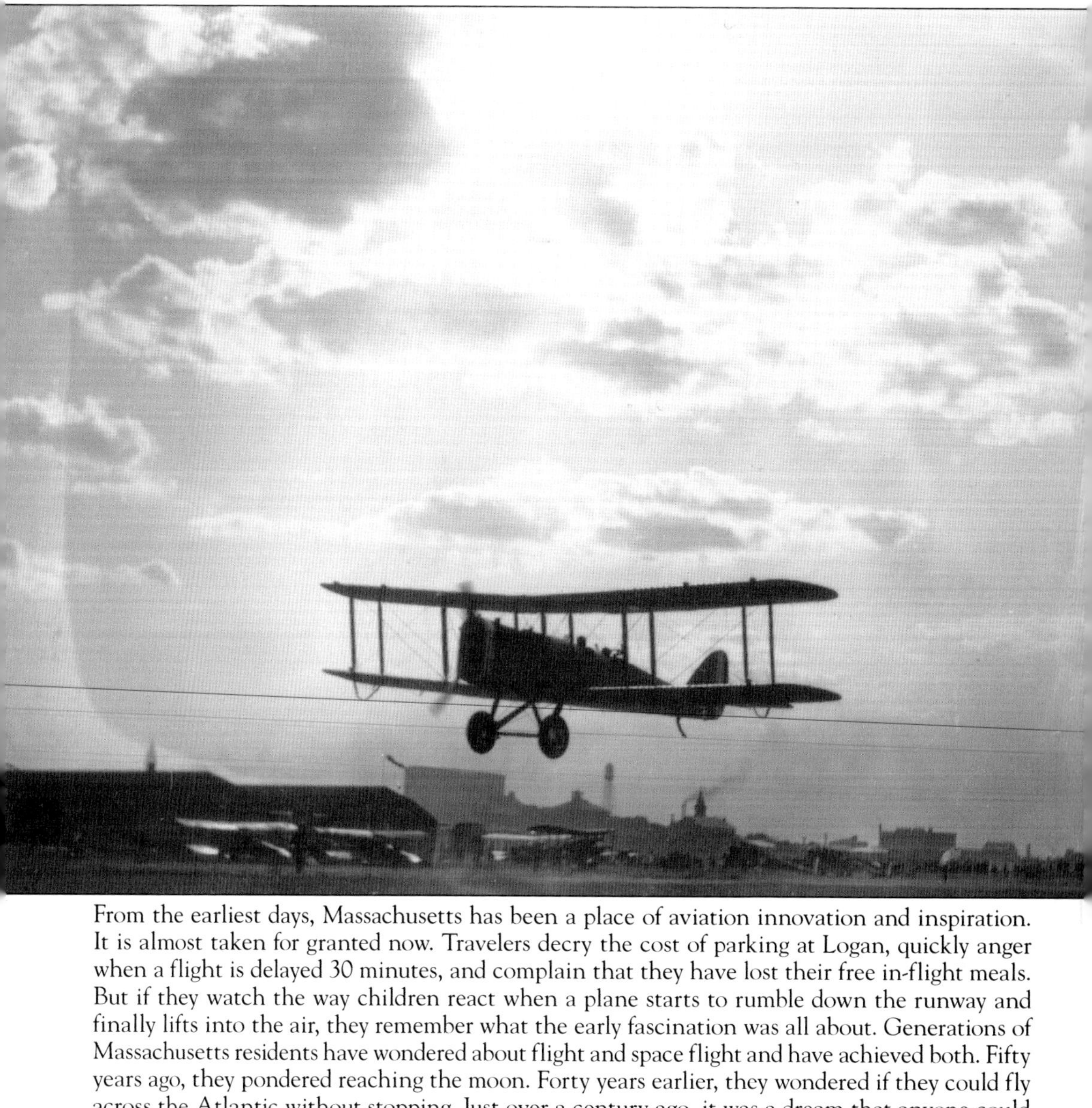

From the earliest days, Massachusetts has been a place of aviation innovation and inspiration. It is almost taken for granted now. Travelers decry the cost of parking at Logan, quickly anger when a flight is delayed 30 minutes, and complain that they have lost their free in-flight meals. But if they watch the way children react when a plane starts to rumble down the runway and finally lifts into the air, they remember what the early fascination was all about. Generations of Massachusetts residents have wondered about flight and space flight and have achieved both. Fifty years ago, they pondered reaching the moon. Forty years earlier, they wondered if they could fly across the Atlantic without stopping. Just over a century ago, it was a dream that anyone could fly. Now, partially thanks to the many Massachusetts aviation pioneers, aircraft designers, engine builders, and more, everyone can. (MAHS.)

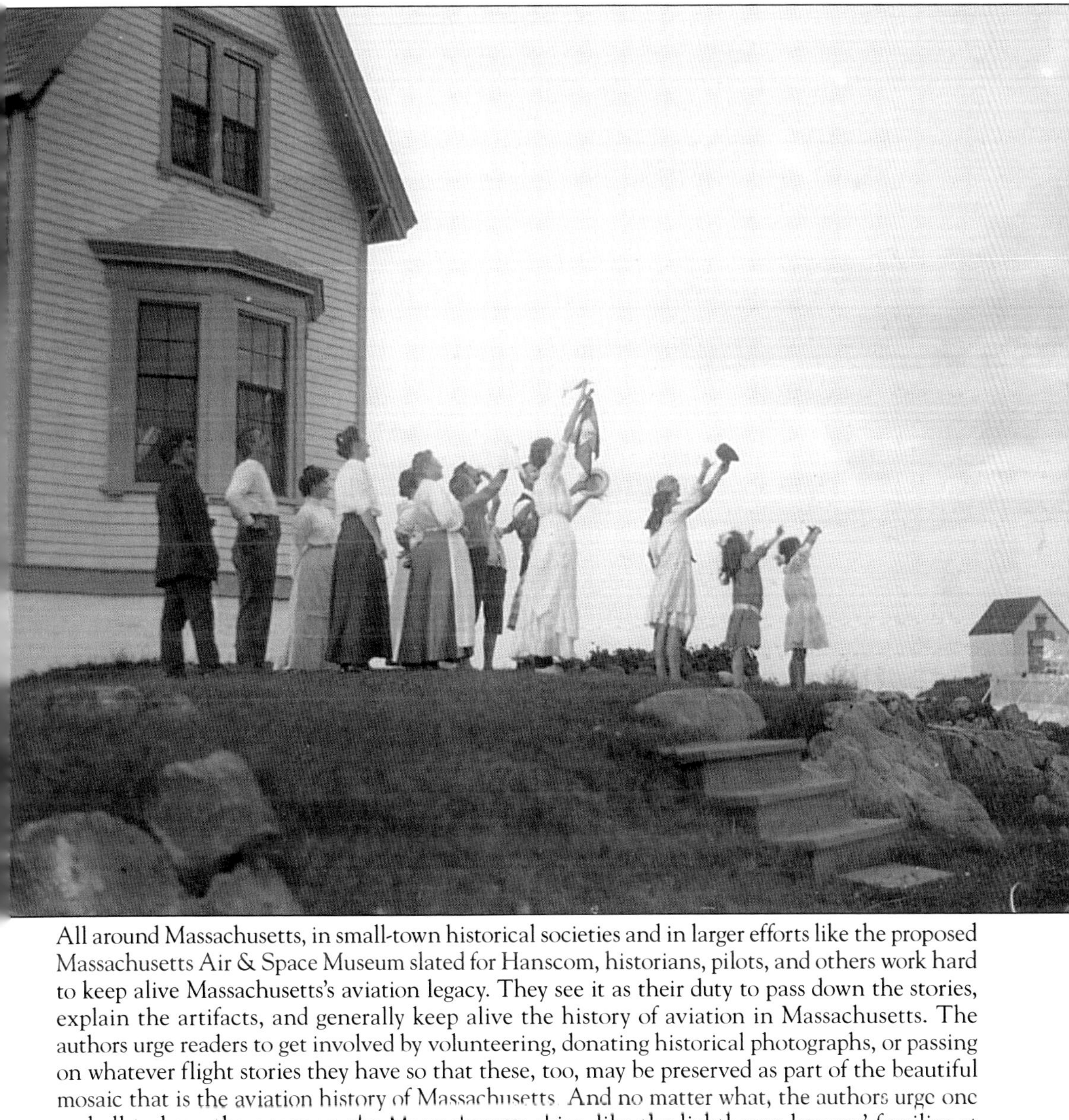

All around Massachusetts, in small-town historical societies and in larger efforts like the proposed Massachusetts Air & Space Museum slated for Hanscom, historians, pilots, and others work hard to keep alive Massachusetts's aviation legacy. They see it as their duty to pass down the stories, explain the artifacts, and generally keep alive the history of aviation in Massachusetts. The authors urge readers to get involved by volunteering, donating historical photographs, or passing on whatever flight stories they have so that these, too, may be preserved as part of the beautiful mosaic that is the aviation history of Massachusetts. And no matter what, the authors urge one and all to keep their eyes on the Massachusetts skies, like the lighthouse keepers' families at Boston Light did when Claude Grahame-White flew overhead in 1910. From balloons to blimps, biplanes to helicopters, jets, rockets, and now drones, Massachusetts has seen it all. What comes next? If we keep our heads down, we may never know. (QHS.)